KT-573-766

Available soon:

For more information visit our website

www.oup.com/vsi/

David Norman

DINOSAURS

A Very Short Introduction
SECOND EDITION

OXFORD

UNIVERSITY PRESS

Great Clarendon Street, Oxford, OX2 6DP,
United Kingdom

Oxford University Press is a department of the University of Oxford.
It furthers the University's objective of excellence in research, scholarship,
and education by publishing worldwide. Oxford is a registered trade mark of
Oxford University Press in the UK and in certain other countries

First edition published 2005
Second edition published 2017

Impression: 3

Published in the United States of America by Oxford University Press
198 Madison Avenue, New York, NY 10016, United States of America

British Library Cataloguing in Publication Data
Data available

Library of Congress Control Number: 2017935370

ISBN 978-0-19-879592-6

Printed in Great Britain by
Ashford Colour Press Ltd., Gosport, Hampshire.

Contents

List of illustrations

The publisher and the author apologize for any errors or omissions in the above list. If contacted they will be pleased to rectify these at the earliest opportunity.

Introduction

Dinosaurs: facts and fiction

Dinosaurs were 'born' officially in 1842 as a result of some truly brilliant and intuitive detective work by the British anatomist Richard Owen, whose work had focused upon the unique features he had identified in some extinct British fossil reptiles.

At the time of Owen's work he was reviewing a surprisingly meagre collection of fossil bones and teeth that had been discovered up to that time and were scattered in private or museum collections around the British Isles. Although the birth of dinosaurs was relatively inauspicious (first appearing as an after-thought in the published report of the eleventh meeting of the British Association for the Advancement of Science), they were soon to become the centre of worldwide attention. The reason for this was simple. Owen worked in London at a time when the influence of the British Empire was arguably at its greatest. To celebrate such influence and industry, the Great Exhibition of 1851 was devised. To house this event a huge but temporary exhibition hall (Joseph Paxton's steel and glass 'Crystal Palace') was built on Hyde Park in central London.

Rather than destroy the Crystal Palace at the end of 1851 it was dismantled and moved to a permanent site in the London suburb of Sydenham (the future Crystal Palace Park). The land surrounding the exhibition building was landscaped and arranged thematically. One of the themes depicted our understanding of natural history and geology, and how these subjects had combined to unravel our Earth's ancient history. This geological park, one of the earliest of its kind, included reconstructions of genuine geological features (caves, limestone pavements, geological strata, and faults) as well as representations of the inhabitants of the ancient world whose fossilized remains had been found in these rocks. Owen collaborated with the sculptor and entrepreneur Benjamin Waterhouse Hawkins to populate the parkland with gigantic, iron-framed, and concrete-clad models of dinosaurs (Figure 1) and other prehistoric creatures known at this time. The advance publicity generated before the relocated Great Exhibition was re-opened in June 1854 included a celebratory dinner held on New Year's Eve 1853 within the belly of a half-completed model of the dinosaur, *Iguanodon*. Coverage of this event in the press excited keen public interest in Owen's dinosaurs.

The fact that dinosaurs ('fearfully great lizards') were extinct denizens of unsuspected earlier worlds, and the embodiment of the dragons of myth and legend, probably guaranteed their adoption by society at large; their names even appeared in the novels of Charles Dickens, a personal acquaintance of Richard Owen. From such evocative beginnings public interest in dinosaurs has been nurtured and maintained ever since. Quite why the appeal should have been so persistent has been much speculated upon: it may have much to do with the importance of story-telling as a means of stimulating human imaginative and creative abilities. It strikes me as no coincidence that in humans the most formative years of intellectual growth and cultural development, between the ages of about 3 and 10 years, are often those when the enthusiasm for dinosaurs is greatest—as many parents can testify. The buzz of excitement created when children glimpse their first dinosaur skeleton is almost palpable. Dinosaurs, as the late Stephen Jay Gould—arguably our greatest popularizer

1. A sketch for *Iguanodon* (top) and one of the completed *Megalosaurus* models constructed for the Crystal Palace Park 'dinosaur island' (bottom).

of scientific natural history—memorably remarked, are popular because they are 'big, scary, and dead'. And it is true that their gaunt skeletons exert a gravitational pull on the imaginative landscape of youngsters—just watch their reaction when glimpsing their very first dinosaur skeleton in a museum...they are transfixed.

A remarkable piece of evidence in support of the notion that there is a relationship between the latent appeal of dinosaurs and the human psyche can be found in mythology and folklore. Adrienne Mayor (a historian at Stanford University) has shown that as early as the 7th century BC the Greeks had contact with nomadic cultures in central Asia. Written accounts at this time include descriptions of the Griffin (or Gryphon): a creature that reputedly hoarded and jealously guarded gold; it was wolf-sized with a beak, wings, four legs with sharp claws on its feet, and a long tail. Furthermore, Near Eastern art of at least 300 BC depicts Griffin-like creatures, as does that of the Mycenaeans. The Griffin myth arose in Mongolia/north-west China, from traders using the ancient caravan routes that visited gold mines in the Tienshan and Altai Mountains. This part of the world (we now know) has a very rich fossil heritage and is notable for the abundance of well-preserved dinosaur skeletons. Skeletal remains are remarkably easy to observe because their white fossilized bones stand out clearly against the soft, red sandstones in which they are buried. Of even greater interest in this context is the fact that the most abundant of the dinosaurs preserved in these sandstones is *Protoceratops*; these are approximately wolf-sized, have a prominent hooked beak, and four legs terminated by sharp-clawed toes. Their skulls also bear striking upswept bony frills behind, which might easily be the origin of the 'wings' that are depicted in Griffin imagery (Figure 2). It seems likely that Griffins owe their origin to genuine observations of dinosaur skeletons made by nomadic travellers passing through Mongolia or north-western China; they demonstrate a remarkable connection between exotic mythological beasts and genuine dinosaurs.

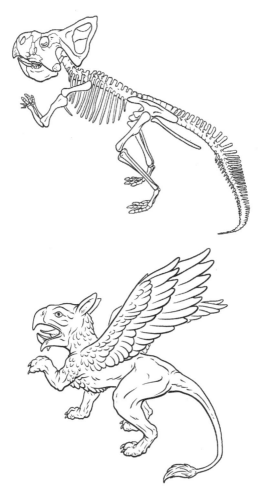

2. The dinosaur *Protoceratops* as the template for the mythological Griffin: a skeletal reconstruction of *Protoceratops* based on numerous skeletons that have been discovered in north-western China and Mongolia (on the ancient Silk Route) (top); an image of the mythological Griffin (bottom). The anatomical similarities between these two images are quite striking.

Looked at through the harsh lens of objectivity, the cultural pervasiveness of dinosaurs is extraordinary. After all, no human being has ever seen a living non-avian dinosaur (no matter what some of the more absurd creationist literature might claim, or what you might see in some movies). The very first recognizably human members of our species lived a mere 500,000 years ago. By contrast, the very last non-avian dinosaurs trod our planet approximately sixty-six million years ago and perished, along with many other creatures, during a cataclysmic event in Earth history (see Chapter 7). Dinosaurs, as a group of animals of quite bewildering variety, are thought to have persisted on Earth for over 170 million years *before* their sudden demise. This surely puts the span of human existence and our current dominance of this fragile planet (in particular, the debates concerning population growth, utilization of resources, pollution, and global warming) into a decidedly sobering perspective.

The very fact of the recognition of dinosaurs, and the very different world in which they lived, today is a testament to the human intellect and the explanatory power of science. The ability to be inquisitive: to probe the natural world and to keep asking that beguilingly simple question—why?—is one of the essences of being human. It is hardly surprising that developing rigorous methods in order to generate plausible answers to our questions is at the core of all science.

Dinosaurs are undeniably interesting to many people. Their very existence and extraordinary range of body forms incite curiosity; this can be used in some instances as a means of introducing unsuspecting audiences to the excitement of scientific discovery and the application and use of science more generally. Just as fascination with bird or whale songs could lead to an interest in the physics of sound transmission, echolocation, auditory physiology, and ultimately applications such as radar, on the one hand, or linguistics and psychology on the other, so it can be that an interest in dinosaurs—their lives, their evolutionary relationships, their behaviours, their locomotor mechanics, or even their metabolic physiologies—open up pathways into a surprising and unexpectedly wide range of scientific disciplines. Outlining some of these pathways across various branches of science is one of the underlying purposes of this book.

Palaeontology is a branch of science that has been built around the study of fossils; technically these are the remains of organisms that died more than 10,000 years ago (that is to say prior to the time when human culture began to have an identifiable impact on the world). The latter is the province of archaeology. Palaeontology attempts to bring fossils back to life: not literally, as in resuscitating dead creatures (as portrayed in the *Jurassic Park* films), but by using science to understand as fully as we can what such creatures were really like, how they fitted into their world, and how that world differs from the present one. When a fossilized animal is discovered its remains present the palaeontologist with a series of tantalizingly incomplete clues, not unlike those faced by the fictional sleuth Sherlock Holmes on his arrival at the scene of a crime (however, in this particular 'case' the skeletal remains are of a *very* long-dead creature). Basic observations and clues can prompt a range of questions:

- What type of creature was it when it was alive?
- How long ago did it die?
- Did it die naturally of old age, or was it killed?
- Did it die just where it was found, buried in the rock, or was its body moved from somewhere else?
- Was it male or female?
- How did the creature appear when it was alive?
- Was its body colourful or drab?
- Was it fast-moving or a slow-coach?
- What did it eat?
- How well could it see, smell, or hear?
- Was it capable of complex behaviour?
- Is it related to any creatures that are alive today?

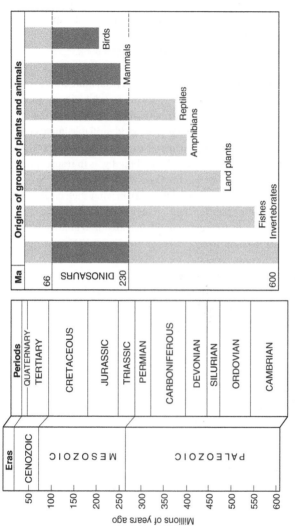

3. The Geological Timescale attempts to put into historical context the time interval (the Mesozoic Era) during which dinosaurs are known to have lived on Earth. Interestingly the birds, still with us today, are direct descendants of one group of dinosaurs—so in a sense it is true to say that some dinosaurs are still with us today, they are just disguised as birds.

8

These are just a few examples of the questions that might be asked, but all tend toward the piecemeal reconstruction of a picture of the creature and of the world in which it lived. It has been my experience, following on from the first broadcasting of the television series called *Walking with Dinosaurs*, with their realistic-looking virtual dinosaurs and the incredible CGI dinosaurs created for the *Jurassic Park* franchise, that many people have been sufficiently intrigued to ask: 'How do we know that they moved like that?...looked like that?...behaved like that?...sounded like that?'

Questions driven by straightforward observation and basic common sense underpin this book. Every fossil discovery is unique and has the potential to teach the inquisitive among us something about our heritage. I should, however, qualify this statement by adding that the particular type of heritage that I will be discussing relates to the *natural* heritage that we share with all other organisms on this planet. This natural heritage spans ~3,800 million years according to most modern estimates. I will be surveying a comparatively small portion of this staggeringly long period of time during which life has existed on Earth: just that interval between ~230 million years ago (MA) and ~66 Ma, when dinosaurs seemed to dominate most aspects of terrestrial life on Earth (see Figure 3).

Chapter 1
Dinosaurs in perspective

The fossilized remains of dinosaurs (with the notable exception of their lineal descendants, birds—Chapter 5) have been found in rocks identified as belonging to the Mesozoic Era. Mesozoic rocks range in age from 252 to 66 Ma (in round figures). In order to put the time during which dinosaurs lived into context, since such numbers are so large as to be quite literally unimaginable, it is easier to refer the reader to a simplified geological timescale—see Figure 3: dinosaurs lived within the shaded zone on the right-hand panel.

During the 19th and a considerable part of the 20th centuries, the age of the Earth, and the relative ages of the different rocks of which it is composed, had been the subject of intense scrutiny. During the early part of the 19th century it was becoming recognized (though not without dispute) that the rocks of the Earth, and the fossils that they contained, could be divided into qualitatively different types. There were rocks that appeared to contain no fossils (often referred to as igneous, or 'basement'). Positioned above these apparently lifeless basement rocks was a sequence of four types of rocks that signified four ages of the Earth. During much of the 19th century these were named Primary, Secondary, Tertiary, and Quaternary—quite literally the first, second, third, and fourth ages of the Earth. The ones that contained traces of ancient shelled and simple fish-like creatures were 'Primary' (now more commonly called *Paleozoic* (also spelled Palaeozoic), meaning 'ancient life'). Above the paleozoics was a sequence of rocks that contained a combination of shells, fish, and land-living *saurians* (or 'crawlers', which today would include animals that we commonly call amphibians and reptiles); these rocks were designated broadly as 'Secondary' (nowadays *Mesozoic*, or 'middle life'). Above the mesozoics were found rocks that contain creatures more similar to those living today, notably because they include mammals and birds; these were named *Cenozoic* (or 'recent life'). And finally, there is the *Quaternary* (or 'fourth age'—sometimes referred to as 'the Recent') that charted the appearance of recognizably modern plants and animals and the influence of the great ice ages. It is also becoming fashionable to refer to a most recent interval of time as the *Anthropocene* (or 'human time'). This covers the last seventy years or so, during which humans have begun to leave measurable traces of their activity on the Earth: plastic pollution, the prevalence of concrete, radioactive debris, the soot generated by power stations, and—more worryingly—the undeniable mass-extinction of organisms.

This general pattern has stood the test of time remarkably well. All modern geological timescales continue to recognize these relatively crude, but fundamental, subdivisions: Paleozoic, Mesozoic, Cenozoic. However, refinements in the way the fossil record can be examined, for example, through the use of high-resolution microscopy, the identification of chemical signatures associated with life, and the more accurate dating of rocks enabled by radioactive isotope techniques have led to greater precision in the dating of the component parts of the timescale of Earth history.

The part of the timescale that we are most concerned with in this book is the Mesozoic Era, comprising three geological Periods: the Triassic (~252–201 Ma), the Jurassic (~201–145 Ma), and the Cretaceous (~145–66 Ma). Clearly these intervals of time are not by any means equal in duration. Geologists were not able to identify a metronome-like tick of a clock measuring the passing of Earth time. The boundaries between the named Periods were established upon the characteristic features of the rocks themselves as well as the distinctive fossils that the rocks contained. The term Triassic originates from a triplet of distinctive rock types found in Germany (known as the Bunter, Muschelkalk, and Keuper); the Jurassic hails from a sequence of rocks identified in the Jura Mountains of France; while the name Cretaceous was chosen to reflect the great thickness of chalk (*Kreta* in Greek) that forms the White Cliffs of Dover and is found widely across Europe, Asia, and North America.

The earliest dinosaur fossils known have been identified in rocks dated at ~230 Ma, from the latter part of the Triassic Period (a Stage named the Carnian), in South America, Africa, and India. Rather disconcertingly, these earliest remains are not rare examples of one type of creature that might be the common ancestor of all later dinosaurs. To date at least three, possibly four, different creatures have been identified: several bipedal meat-eaters (*Eoraptor* and *Staurikosaurus*), a tantalizingly incomplete plant-eater named *Pisanosaurus*, and an as-yet-unnamed omnivore. One conclusion seems obvious: these cannot be the *first* dinosaurs. In the Carnian there was already a diversity of early dinosaurs (or dinosaur-like animals such as *Herrerasaurus* and its near relatives). This implies that there must have been dinosaurs living in the Middle Triassic (the Ladinian-Anisian stages) that had 'fathered' the Carnian diversity. Indeed dinosaur-like footprints (and a few rather scrappy bones named *Nyasasaurus*) of possibly Anisian age have been described, but rather frustratingly the bones are scattered and difficult to interpret (they might even belong to more than one animal); and we cannot know what the animal that left the tracks actually looked like (except that it was small and had dinosaur-like feet). So we know for a fact that the story of dinosaur origins, both the time and the place, is incomplete and this stimulates on-going research.

Why dinosaur fossils are rare

It is important, at the outset, for the reader to realize that the fossil record is incomplete and, perhaps more worryingly, decidedly patchy. Incompleteness is an inevitable consequence of the process of fossilization. Dinosaurs were all land-living (terrestrial) animals; this creates particular difficulties. To appreciate what these are it is necessary first to consider the case of a shelled creature such as an oyster living on the sea floor. In the shallow seas where oysters live today, their *fossilization potential* is quite high. They are living on, and attached to, the seabed and are subjected to a constant 'drizzle' of small particles (sediment), including the shells of decaying planktonic organisms, silt, or mud and sand grains. If an oyster dies, its soft tissues will rot or be scavenged by other organisms quite quickly but its hard shell will be gradually buried under layers of sediment. Once buried, the shell has the potential to become a fossil as it becomes trapped under increasingly thick layers of sediment. Over thousands or millions of years, the sediment in which the shell has been buried is gradually compressed and this may become cemented or lithified (literally, 'turned to stone') by the deposition of calcium carbonate (calcite/chalk) or silica (chert/flint) derived from dissolved minerals that are carried through the fabric of the rock by percolating water. For the trapped fossil remains of the oyster to be discovered, the buried rock would need to be lifted by earth movements to form dry land and then subjected to the normal processes of weathering and erosion.

Land-living creatures, by contrast, have a far lower probability of becoming fossilized. Any animal dying on land is most likely to have its soft, fleshy tissues decay or scavenged; to be preserved as a fossil the remains would need to be subject to some form of burial (as with the oyster). In very rare circumstances creatures may be buried rapidly in drifting dune sand, a mud-slide, or even under volcanic ash. However, in the majority of cases the remains of land animals would need to be washed into a nearby stream or river, and eventually find their way into a lake or seabed for the process of slow burial, leading to fossilization, to commence. In simple probabilistic terms, the pathway leading to fossilization of the remains of any land creature is longer and fraught with far greater hazard. Many animals that die on land are scavenged and their remains become entirely scattered and disintegrate so that even skeletal hard parts are ground to dust and recycled back into the geosphere; others have their skeletons scattered, so that only broken fragments actually complete the path to eventual burial, leaving tantalizing (and puzzling) glimpses of creatures—only very rarely will major parts, or even whole skeletons, be preserved.

So, logic dictates that dinosaur skeletons (as with those of any land-living animal) should be extremely rare—and indeed they are, despite the impression sometimes given by the media.

The discovery of dinosaurs and their appearance within the fossil record is also a decidedly chancy business, for rather mundane reasons. Fossil preservation is, as we have just come to appreciate, a haphazard rather than design-driven process. The actual discovery of fossils is similarly serendipitous in the sense that outcrops of rocks are not neatly arranged like the pages of a book to be sampled in sequence or as fancy takes us.

The relatively brittle surface layers of the Earth (its Crust, in geological terms) have been torn or buckled by huge geological forces acting over tens or hundreds of millions of years. As a result, the geological strata containing fossils have been broken, thrown up, and frequently destroyed completely throughout geological time. What we, as palaeontologists, are left with is an extremely complex 'battlefield', pitted, cratered, and broken in a bewildering variety of ways. Disentangling this 'mess' has been the work of countless generations of field geologists. Outcrops here, cliff-sections there, have been studied and slowly assembled into the jigsaw that is the geological structure of the land.

As a result, it is now possible to identify rocks of Mesozoic age (belonging to the Triassic, Jurassic, and Cretaceous Periods) with some accuracy in any country in the world. However, that is not sufficient to aid the search for dinosaurs. It is also necessary to disregard Mesozoic rocks laid down on the sea floor, such as the thick chalk deposits of the Cretaceous and the abundant marine limestones of the Jurassic. The best types of rocks to search in for dinosaur fossils are those that were laid down in shallow coastal or estuarine environments; these might have trapped the odd, bloated carcasses of land-living creatures washed out to sea. But best of all are river and lake sediments, environments that were physically much closer to the source of the majority of land creatures.

Searching for dinosaurs

From the very outset, we need to approach the search for dinosaurs systematically. On the basis of what we have learned so far, it is first necessary to check where to find rocks of the appropriate age by consulting geological maps of the country that is of interest. It is equally important to ensure that the rocks are of a type that is at least likely to preserve the remains of land animals; so some geological knowledge is required in order to predict the likelihood of finding dinosaur fossils, especially when visiting an area for the first time.

Mostly, this involves developing a familiarity with rocks and their appearance in the area being investigated; this is rather similar to the way in which a hunter needs to study intently the terrain in which their prey lives. It also requires the development of an 'eye' for fossils, which comes simply from observing carefully until you begin to recognize particular colours and textures, but this takes time.

Discovery provides an adrenaline-rush of excitement, but it is also the time when the discoverer needs to be most circumspect. All too often fossil discoveries have been ruined, scientifically speaking, in a frantic bid to dig the specimen up, so that the proud finder can display it. Such impatience can result in great damage to the fossil itself—worse still if the object is revealed to be part of a larger skeleton that might have been far more profitably excavated by a larger team of trained palaeontologists. And, as can be imagined, the rocks in which the fossil was embedded may also have important tales to tell concerning the circumstances under which the animal died and was buried, in addition to the more obvious information concerning the actual geological age of the specimen.

The search for fossils can be an exciting adventure as well as a technically fascinating process. However, finding fossils is just the beginning of a process of scientific investigation that can lead to an understanding of the biology and way of life of the fossilized creature and the world in which it once lived. In this latter respect, the science of palaeontology exhibits some similarities to the work of the forensic pathologist: both clearly share an intense interest in understanding the circumstances surrounding the discovery of a body, or its fragmentary remains, and use scientific techniques to understand and interpret as many of the remaining clues as possible.

Dinosaur discovery: *Iguanodon*

Having found a fossil, it needs to be studied scientifically in order to reveal its identity, its relationship to other already known organisms, as well as more detailed aspects of its appearance, biology, and ecology. To illustrate a few of the trials and tribulations inherent in any such programme of palaeontological investigation, we will examine a rather familiar and well-studied dinosaur: *Iguanodon* (Figure 1). This dinosaur has been chosen because it has an interesting and relevant story to tell, and one with which I am particularly familiar, because it proved to be the unexpected starting point for my career as a palaeontologist. Serendipity seems to have a significant role to play in palaeontology and this is certainly true for my own work.

The story of *Iguanodon* spans the entire history of scientific research on dinosaurs and almost the entire history of the science of palaeontology. As a result, this animal unwittingly illustrates the progress of scientific investigation on dinosaurs (and other areas of palaeontology) over the past 200 years. The story also reveals scientists as human beings with passions and struggles, and the pervasive influence of 'pet theories' at various times in the history of the subject.

The first *bona fide* records of the fossil bones that were later to be named *Iguanodon* date back to 1809. They comprise, among indeterminable broken fragments of vertebrae, the lower half of a distinctive tibia (shin bone) collected from a quarry at Cuckfield in Sussex. William Smith (often referred to as the 'father of English geology') collected these fossils. At the time, Smith was preparing the first geological map of Britain, which he completed in 1815. Although these fossil bones were clearly sufficiently interesting to have been collected and preserved (they are still in the collections of the Natural History Museum, London) their meaning was not understood. The fossil languished, forgotten by science, until I studied them in the late 1970s.

1809 was a remarkably opportune moment for such a discovery to be made. Things were happening in Europe in the branch of science concerned with fossils. One of the greatest and most influential scientists of this age (Georges Cuvier, 1769–1832) was a *naturalist* working in Paris (and an administrator in the Emperor Napoleon's government). Naturalist was the name applied to a philosopher-scientist who worked on a wide range of subjects associated with the natural world: the Earth, its rocks and minerals, fossils, and all living organisms. In 1808, Cuvier re-described a renowned gigantic fossil skull collected from a chalk quarry at Maastricht in Holland; its renown stemmed from the fact that it had been claimed as a trophy of war by Napoleon's army during the siege of Maastricht in 1795. Cuvier identified the skull as that of an enormous marine lizard (named

Mosasaurus by William Daniel Conybeare). The effect of this revelation—the existence of an unexpectedly gigantic fossil lizard in an earlier time in Earth history—was profound. It encouraged the search for other giant fossils; and it established, beyond reasonable doubt, that pre-biblical 'earlier worlds' had existed. It also determined a particular way of viewing and interpreting such fossil creatures—as gigantic lizards.

Following the defeat of Napoleon (1815) and the restoration of peace between England and France, Cuvier was able to visit England in 1817–18 and meet scientists with similar interests. At Oxford he was shown some gigantic fossil bones in the collections of the geologist William Buckland; these seemed to belong to an enormous, land-living, lizard-like creature. Buckland (with a little help from Conybeare) described and named this creature *Megalosaurus* in 1824.

From the perspective of our story, the really important discoveries were not made until around 1821–2 and at the same quarries, around Whiteman's Green in Cuckfield, visited by William Smith some twelve years earlier. At this time an energetic and ambitious medical doctor (Gideon Mantell, 1790–1852), living in the town of Lewes, was dedicating all his spare time to completing a detailed report on the geological structure and fossils in his native Weald district (an area incorporating much of Surrey, Sussex, Kent, and a little bit of Hampshire) in southern England.

Mantell's work culminated in an impressively large, well-illustrated book that he published in 1822. He included descriptions of several large reptilian teeth (Figure 4) and broken ribs that he was unable to classify.

Mantell purchased these fossil remains from quarrymen, and his wife (Mary Ann) collected others herself. The next three years

4. *Iguanodon* teeth. These are two fine examples of the teeth described by Gideon Mantell and named *Iguanodon* in 1825. On the left is an upper (maxillary) tooth, with its tip abraded by feeding on plants. On the right is a lower (dentary) tooth similarly worn at its tip. Both teeth are embedded in pieces of Tilgate Grit (a rock type excavated at Cuckfield Quarry).

saw Mantell struggling to identify the type of animal to which such large fossil teeth might have belonged. Although not trained in comparative anatomy (the particular specialism of Cuvier), he developed contacts with many learned men in England in the hope of gaining some insight. He also sent some of his precious specimens to Cuvier in Paris for an expert opinion. Cuvier at first dismissed Mantell's discoveries as fragments of recent animals: perhaps the incisor teeth of a rhinoceros, or even those of large, coral-chewing, bony fish—were Cuvier's suggestions. Undeterred, Mantell continued to investigate and finally found a solution. In the collections of the Royal College of Surgeons in London he was shown the skeleton of an iguana, a large herbivorous lizard that had just been discovered in South America. Its teeth were similar in general shape to those of his fossils (but far smaller) and this suggested to Mantell that his fossils belonged to a gigantic, extinct relative of the iguana. In 1825 Mantell published a report on his discoveries and the name chosen was *Iguanodon* ('iguana tooth'). The name was created (yet again!) at the suggestion of Conybeare (the latter clearly had a knack for inventing euphonious names for these early discoveries).

These early discoveries reinforced the idea of an ancient world inhabited by improbably large lizards. A simple scaling of the minute teeth of the living (1 metre-long) iguana with those of Mantell's *Iguanodon* suggested a body length in excess of 25 metres! The excitement and personal fame generated by his description of *Iguanodon* drove Mantell to greater efforts to discover more about this new animal and other fossil inhabitants of the ancient Weald.

For several years after 1825 only tantalizing fragments of Weald fossils were discovered; then, in 1834, a partial, disarticulated skeleton (Figure 5) was discovered at a quarry near Maidstone in Kent. Eventually purchased from the quarry owner for Mantell by some of the latter's friends, it was christened the 'Mantel-piece' and inspired much of his later work. It was the basis for the first visualization of a dinosaur skeleton (Figure 6). He continued probing the anatomy and biology of *Iguanodon* in his later years,but much of this was, alas, overshadowed by the rise of an extremely able, well-connected, ambitious, and ruthless personal nemesis: Richard Owen (1804–92).

5. The 'Mantel-piece': a photograph of the 'Mantel-piece' as it appears today in the collections of the Natural History Museum, London (top); a sketch based upon the photograph above with all the bones identified by the author (bottom). This partial skeleton (it is surprisingly complete) was discovered in 1834 during rock-blasting at a quarry near Maidstone in Kent.

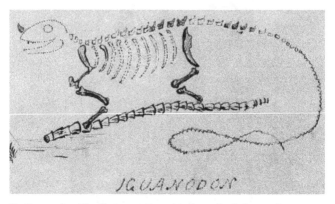

6. *Iguanodon.* The first ever attempt to draw the skeleton of a dinosaur (*c.*1834). Mantell drew this using the evidence of the bones preserved on the 'Mantel-piece'. The darker shaded bones are some of the actual fossil bones that can be seen clearly on the original slab.

The 'invention' of dinosaurs

Fourteen years younger than Mantell, Richard Owen also studied medicine, but concentrated in particular on anatomy. He gained a reputation as a skilled anatomist, and acquired a position at the Royal College of Surgeons in London. In due course, he became known as the 'English Cuvier'. During the late 1830s, Owen persuaded the British Association to grant him money to prepare a detailed review of all known British fossil reptiles. This resulted in the production of a stream of large, well-illustrated books that mimicked Cuvier's multi-volume, *Ossemens Fossiles*.

Owen's first productions were a report in 1840 on marine fossils (Conybeare's *Enaliosauria*) and another in 1842 on the remainder, including Mantell's *Iguanodon*. The 1842 report is a remarkable document because in it Owen writes of a new 'tribe or sub-order…which I…name…Dinosauria'. Owen identified three dinosaurs in this report: *Iguanodon* and *Hylaeosaurus*, both discovered in the Weald and named by Mantell, as well as *Megalosaurus*, the giant reptile from Oxford. He recognized

No. 7. Megalosaurus.

7. *Megalosaurus.* **This sketch was probably made under Owen's guidance in preparation for the construction of the Crystal Palace models (*c.*1853). It clearly shows the then known bones of *Megalosaurus* and how Owen thought they were arranged, and how he thought that dinosaurs might have looked in real life. This was radically different to Mantell's vision.**

dinosaurs as a unique and previously unrecognized group on the basis of several distinctive anatomical observations. These included the *enlarged sacrum* (a remarkably strong attachment of the hips to the spinal column), the *double-headed ribs* in the chest region, and *pillar-like legs* (Figure 7).

Owen trimmed their dimensions considerably, suggesting that they were large, but in the region of 9 to 12 metres long, rather than the more dramatic lengths suggested by Cuvier, Buckland, and Mantell. Furthermore, Owen speculated a little more on the anatomy and biology of these animals in words that have extraordinary resonance in the light of today's understanding of the biology and way of life of dinosaurs.

Among his concluding remarks in the report, he observed that dinosaurs:

> attained the greatest bulk, and must have played the most
> conspicuous parts, in their respective characters as devourers of

animals and feeders upon vegetables, that this earth has ever
witnessed in oviparous [egg-laying] and cold-blooded creatures.
(Owen 1842, p.200)

And a little later:

The Dinosaurs having the same thoracic structure [double-headed
ribs] as the Crocodile, may be concluded to have possessed a
four-chambered heart...more nearly approaching that which now
characterizes the warm-blooded Mammalia. (Owen 1842, p.204)

Owen imagined them to be stout, egg-laying, and scaly
creatures (because they were still reptiles) that resembled the
largest living mammals found in the tropical regions today.
His dinosaurs were the 'crowning glory' of a time on Earth
when egg-laying and scaly-skinned reptiles reigned supreme.
As a piece of logical deduction based on such meagre remains
(the known bones are inked in Figure 7), this was Owen at
his most brilliant. Such breath-taking vision is even more
remarkable when set against the 'gigantic lizard' models
of the great Cuvier.

But Owen's creation of the Dinosauria had another important
purpose. His reports offered a scientific refutation of the
progressionist and *transmutationist* movements within the
fields of biology and geology during the first half of the 19th
century. Progressionists noted that the fossil record seemed to
show that life *progressed*: the earliest rocks revealed simple
forms of life, whereas younger rocks contained more complex
creatures. Transmutationists noted that members of one species
were not identical and wondered whether this variability might
permit species to change over time (or *evolve*—as we would say
today). Jean Baptiste de Lamarck, a colleague of Cuvier in Paris,
had suggested that animal species might 'transmute' (evolve)
over time through his theory of the inheritance of acquired
characteristics. Such ideas challenged the widely held, biblically

inspired, belief that God had created all creatures on Earth. Such revolutionary ideas were fiercely debated at this time.

Dinosaurs, and indeed several of the groups of organisms recognized in the God-fearing Owen's reports, provided him with evidence that life on Earth did not show continuous *progress*—in fact he suggested quite the reverse. Dinosaurs were anatomically reptiles (egg-laying, cold-blooded, scaly vertebrates). However reptiles today were, to Owen's mind, a *degenerate* group compared to the 'magnificent' dinosaurs that had lived during Secondary (Mesozoic) times. In short, Owen was attempting to strangle the radical intellectuals of the time; the latter were open to the possibility of 'transmutational' (evolutionary) change and strove for a mechanism that might make this possible. Owen simply wanted to re-establish an understanding of the diversity of life that had its basis closer to the views espoused by Reverend William Paley in his book entitled *Natural Theology*, in which God held centre-stage as the Creator and Architect of all Nature's creatures.

Owen's fame grew steadily through the 1840s and 1850s, and he became involved in the committees associated with the planning for the re-opening of the Great Exhibition in 1854. It is curious that Owen, for all his fame, was not first choice as the scientific director of the construction of the dinosaurs—Gideon Mantell was! Mantell refused to take the position because of persistent ill-health, and because he was wary of the risk of misrepresenting imperfectly developed ideas, such as the body shape of dinosaurs. Owen, not surprisingly, had no misgivings of this sort.

Mantell's story ended in tragedy: his obsession with fossils and the development of a private fossil museum led to the collapse of his medical practice, and his family disintegrated (his wife left him and his surviving children emigrated once they were old enough to leave home). The diary that he kept for much of his life makes melancholy reading. In his final years Mantell was alone and racked by chronic back pain (scoliosis), he eventually died of a self-administered overdose of laudanum.

Although outflanked by the ambitious, brilliant, and (crucially) full-time scientist Owen, Mantell spent the last decade of his life researching 'his' *Iguanodon*. He produced a series of scientific articles and extremely popular books summarizing many of his new discoveries, and was the first to realize (in 1851) that Owen's vision of the dinosaurs (or at least *Iguanodon*) as stout 'elephantine reptiles' was probably incorrect. Analysis of the 'Mantel-piece' revealed that *Iguanodon* had strong back legs and smaller, weaker front limbs. He concluded that its posture may have had much more in common with the 'upright' reconstructions of giant ground sloths (based upon Owen's detailed description of the fossil ground sloth, *Mylodon*). Sadly, Mantell's work was overlooked because of the excitement and publicity surrounding Owen's Crystal Palace dinosaur models. The validity of Mantell's suspicions about the posture of *Iguanodon* and the strength of this man's intellect were not to be revealed for a further thirty years, and then through an amazing piece of serendipity.

Reconstructing *Iguanodon*

In 1878 fossils were discovered in a coal mine in the small village of Bernissart in Belgium. The colliers, who were mining a coal seam over 300 metres beneath the surface, unexpectedly struck shale (laminated clay) in which they began to find large fossilized 'logs'; these were eagerly collected because their interiors seemed to be packed with gold! On closer inspection, the 'logs' turned out to be fossil bones and the gold was 'fool's gold' (angular crystals of iron pyrites). Some fossilized teeth were also discovered among the bones and these were similar to those described as *Iguanodon* by Mantell many years earlier. The miners had discovered not gold, but a veritable treasure-trove of dinosaur skeletons.

Over the next five years, a team of miners and scientists from the Royal Belgian Museum of Natural History in Brussels (Royal Belgian Institute of Natural Sciences) excavated nearly forty skeletons of *Iguanodon*, as well as a great number of animals and plants preserved with the dinosaurs. Many *Iguanodon* skeletons were complete and fully articulated; they represented the most spectacular discovery that had been made anywhere in the world at the time. It was the good fortune of a young scientist in Brussels (Louis Dollo, 1857–1931) to be allowed to study and describe these extraordinary riches, and this he did from 1882 until his retirement in the 1920s.

The complete dinosaur skeletons unearthed at Bernissart showed vividly that Owen's model of dinosaurs such as *Iguanodon* (Figure 1) was incorrect. As Mantell suspected, the front limbs were not as large and strong as the back legs and the animal had a massive tail (Figure 8), so that its proportions more closely resembled those of a gigantic kangaroo.

The new skeletal restoration, and the process by which it was arrived at, are particularly revealing because they show the influence of contemporary ideas concerning the appearance and affinities of dinosaurs. Owen's 'elephantine reptile' vision of the dinosaur had been questioned as early as 1859 by some incomplete dinosaur discoveries made in New Jersey (USA) and studied by Joseph Leidy, a man of equivalent scientific stature to Owen, working at the Philadelphia Academy of Natural Sciences. However, Owen was to be far more roundly criticized by a younger, London-based, and ambitious rival (Thomas Henry Huxley, 1825–95).

By the late 1860s new discoveries had been made that added to the debate over the relationships of dinosaurs to other animals. The first skeleton of the earliest well-preserved fossil bird *Archaeopteryx* ('ancient wing') had been discovered in southern Germany (Figure 9). The specimen was unusual in that it had well-preserved

8. *Iguanodon.* An outline drawing of one of the complete, mounted dinosaur skeletons that are on display at the Royal Belgian Institute of Natural Sciences, Parc Léopold, Brussels.

impressions of feathers, the key identifier for any bird, forming a halo in the matrix around its skeleton. However, unlike any living bird, and disconcertingly similar to modern reptiles, it also had three long, clawed fingers on each foot (birds don't have claws on their feet), teeth in its jaws (rather than a beak), and a long bony tail. Some living birds might seem to have long tails but this is just the profile of their feathers that are anchored in a very short, stump-like remnant of the tail—the 'Parson's Nose'.

Not long after the discovery of *Archaeopteryx*, another small, delicate skeleton was found in the same quarries in Germany (Figure 10). There were no feather impressions (its arms were far too short to have served as wings) and anatomically it was clearly a small, predatory dinosaur. It was named *Compsognathus* ('pretty jaw').

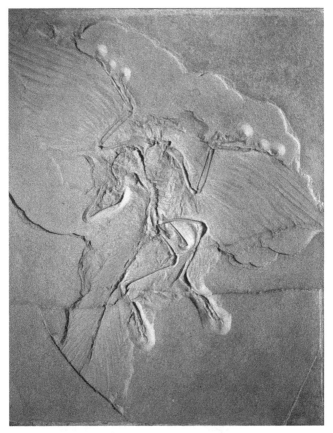

9. *Archaeopteryx*. The remarkably well-preserved skeleton of a small (40 cm long) complete late Jurassic bird-like creature with clear indication of wing and tail feathers preserved as impressions in the fine-grained limey mud (lithographic limestone) in which this creature was buried. This specimen is in the Humboldt Museum of Natural History, Berlin.

10 cm

10. *Compsognathus*. The original fossil of this dinosaur (70 cm long) was preserved on a slab of lithographic limestone collected from a quarry in Bavaria in the 1860s. This reconstruction based upon the original specimen shows the delicate, almost bird-like, nature of the skeleton of this dinosaur.

These two discoveries emerged in timely fashion, scientifically speaking. In 1859, just a year before the first remains of *Archaeopteryx* was unearthed, Charles Darwin published a book entitled *On the Origin of Species*. This book provided a detailed review of the evidence supporting the progressionist and transmutationist speculations mentioned earlier. Most importantly, Darwin suggested a new mechanism—*Natural Selection*—by which transmutation (*evolution*) might occur and produce new species. The book was sensational because it offered a carefully argued challenge to the authority of biblical teachings: that God created all the species on Earth. Darwin's proposal was opposed by pious establishment figures (notably Richard Owen). In contrast, radical (open-minded) intellectuals reacted positively to Darwin's ideas. Thomas Huxley, one of those radicals, declared, 'How extremely stupid not to have thought of that!' (Huxley, 1887, p.197), once he understood Darwin's theory.

While not wishing to become too involved in Darwinian matters here, Huxley was quick to realize that *Archaeopteryx* and the small predatory dinosaur *Compsognathus* were anatomically very similar. By the early 1870s, Huxley was proposing that birds and dinosaurs were not only anatomically similar, but that, based on this evidence, birds had *evolved from* dinosaurs. Louis Dollo, as a bright young student, would have been fully aware of the Owen versus Huxley–Darwin feuds. One burning question must have been: did these new discoveries have any bearing on the great scientific controversy of the day?

Careful anatomical study of the full skeleton of *Iguanodon* (Figure 8) revealed a hip structure that was bird-like; furthermore, it had long back legs that ended in bird-like three-toed feet (similar to those of big, running birds such as emus, cassowaries, and rheas). *Iguanodon* also had a rather swan-like curved neck, and the tips of its jaws were toothless and seemingly bore a bird-like beak. Given the task of description and interpretation faced by Dollo in the immediate aftermath of these spectacular discoveries it is intriguing to note that early photographs of the reconstruction of the first skeleton in Brussels (Figure 11), show two Australian creatures: a wallaby (a small variety of kangaroo) and a cassowary (the large flightless bird).

The influence of the debates raging in England cannot be doubted. This new discovery pointed to the truth implicit in Huxley's arguments (and made it clear that Mantell had been on the right track in 1851). *Iguanodon* was no lumbering, scaly rhinoceros lookalike as portrayed by Owen (Figure 1), it was a huge creature with a pose similar to that of a resting kangaroo, but with a number of remarkably bird-like attributes.

Dollo was tirelessly inventive in his approach to the fossil creatures that he described—he dissected modern lizards, crocodiles, and birds in order to better understand their biology and musculature to see if the information could help him reconstruct the soft tissues of his dinosaurs. In many respects, he was adopting a forensic pathology-like approach to understand his puzzling dinosaurs. Dollo was regarded as the architect of a new style of palaeontology: *palaeobiology* (named in honour of his work) that linked fossils with modern biology, ecology, and behaviour.

Dollo's final contribution was a paper published in 1923 to mark the approximate centenary of Mantell's original discovery of *Iguanodon*. He summarized his views on the dinosaur, identifying it as the dinosaurian ecological equivalent of an arboreal browser similar in some respects to a giraffe. Dollo showed that its posture enabled it to reach high into trees to gather its fodder, which it was able to draw into its mouth by using a long, muscular tongue; its sharp beak was used to nip off tough stems, while its teeth pulped the food before it was swallowed. So firmly was this interpretation adopted, that it stood, literally and metaphorically, unchallenged for the next sixty years. His views were reinforced by the distribution of replica skeletons of *Iguanodon* (by command of King Leopold II)

11. *Iguanodon* being reconstructed. The first partial skeleton recovered from Bernissart is photographed in the early 1880s inside the *Chapelle de Nassau* close to the Royal Museum of Natural History in Brussels. Note the skeletons of a cassowary and a wallaby that were used to aid the reconstruction.

to many of the great museums around the world during the early years of the 20th century, and by many popular and influential textbooks written on the subject.

Dinosaur palaeontology in decline

Paradoxically the culmination of Dollo's remarkable work on this dinosaur, and his international recognition as the 'father' palaeobiology, in the 1920s, marked the beginning of a decline in the perceived relevance of this area of research within the natural sciences.

In the interval between the mid-1920s and the mid-1960s, palaeontology, and particularly the study of dinosaurs, stagnated. The excitement of the early discoveries, notably those in Europe had been succeeded by spectacular 'bone wars' that gripped America during the last three decades of the 19th century. These centred on a furious—and sometimes violent—race to discover and name new dinosaurs, and had all the hallmarks of the 'Wild West'. At its centre were Edward Drinker Cope (a protégé of the polite and unassuming Professor Leidy) and his 'opponent' Othniel Charles Marsh at Yale University. They hired gangs of thugs to venture out into the American mid-West to collect as many new dinosaur bones as possible. This 'war' resulted in a frenzy of scientific publications naming dozens of new dinosaurs, many of whose names still resonate today: *Brontosaurus, Stegosaurus, Triceratops,* and *Diplodocus*.

Equally fascinating discoveries were made, partly by accident, during the early 20th century in exotic places such as Mongolia by Roy Chapman Andrews of the American Museum of Natural History in New York (the real-life hero/explorer upon whom was based the mythical 'Indiana Jones'); and in German East Africa (Tanzania) by Werner Janensch of the Berlin Museum of Natural History.

More new dinosaurs were continually being discovered and named from various places around the world. Although they created dramatic centre-pieces in museums, palaeontologists seemed to be doing little more than adding new names to the roster of extinct creatures. A sense of failure took hold to the extent that some even used dinosaurs as examples of a theory of extinction based on 'racial senescence'. The general idea was that they had lived for so long that their genetic constitution was simply exhausted and no longer capable of generating the novelty necessary for the group as a whole to survive. This supported the view that dinosaurs were a glorious failure: a self-evidently unsuccessful experiment in animal design and evolution.

Not surprisingly, many biologists and theoreticians viewed this area of research with a jaundiced eye. New discoveries, though undeniably exciting, did not seem to be providing data that led in any particular direction. Discovery required the scientific formalities of description and naming of these dead creatures, but beyond that all interest seemed essentially museological. To be brutal, the work was seen as the equivalent of 'stamp collecting'. Dinosaurs, and many other fossil discoveries, offered glimpses of the tapestry of life within the fossil record, but beyond that their scientific value seemed questionable.

Many factors justified this change in perception: Gregor Mendel's work (published in 1866, but overlooked until 1900) on the laws of particulate inheritance (genetics) provided the crucial mechanism to support Darwin's theory of evolution by means of Natural Selection. Mendel's work was elegantly merged with Darwin's theory to create *Neodarwinism* ('new Darwinism') in the 1930s. At a stroke, Mendelian genetics solved one of Darwin's most fundamental worries about his theory: how favourable characteristics (genes or alleles in the new Mendelian language) could be passed from generation to generation. In the absence of any better understanding of the mechanism of inheritance in the mid-19th century, Darwin had assumed that characters or traits, the features subject to Natural Selection according to his theory, were blended when passed to the next generation. This, however, was a fatal flaw, because Darwin realized that any favourable traits would simply be diluted out of existence if they were blended (halved) from generation to generation. Neodarwinism clarified matters enormously, Mendelian genetics provided mathematical rigour to the theory, and the revitalized subject spawned new avenues of research. It created the sciences of genetics and molecular biology, culminating in Crick and Watson's model of DNA in 1953, as well as huge developments in the fields of behavioural evolution and evolutionary ecology.

Unfortunately, this fertile intellectual ground was not so obviously available to palaeontologists. Self-evidently, genetic mechanisms could not be studied in fossil creatures, so it seemed that they could offer no material evidence to the intellectual thrust of evolutionary studies during much of the remainder of the 20th century. Darwin had already foreseen the limitations of palaeontology in the context of his new theory. Using his inimitable reasoning, he noted the limited contribution that could be made by fossils to any of the debates concerning his new evolutionary theory. In a chapter of the *Origin of Species* devoted to the subject of the 'imperfections of the fossil record', Darwin noted that although fossils provided material proof of evolution during the history of life on Earth (harking back to the older progressionists' arguments), the geological succession of rocks, and the fossil record contained within in it, was lamentably incomplete. Comparing the geological record to a book charting the history of life on Earth, he wrote:

> of this volume, only here and there a short chapter has been preserved; and of each page, only here and there a few lines. (Darwin, 1882, 6th edn, p.318)

Dinosaur palaeobiology: a new beginning

It was not until the 1960s and early 1970s that the study of fossils began to re-emerge as a subject of wider and more general interest. The catalyst for this re-awakening was a younger generation of evolutionarily minded scientists eager to demonstrate that the evidence from the fossil record was far from being a Darwinian 'closed book'. The premise that underpinned this new work was that while evolutionary biologists are obviously constrained by working with living animals in an essentially two-dimensional world—they are able to study species, but they do not witness the emergence of new species—palaeobiologists, by contrast, work in the third dimension of time. The fossil record provides sufficient time to allow new species to appear and others to become extinct. This permits palaeobiologists to pose questions that bear on the problems of evolution: does the geological timescale offer an added (or different) perspective on the process of evolution? And, is the fossil record sufficiently informative that it can be teased apart to reveal some evolutionary secrets?

Detailed surveys of the geological record began to demonstrate rich successions of fossils (particularly shelled marine creatures)—considerably richer than Charles Darwin could ever have imagined, given the comparative infancy of palaeontological work in the middle of the 19th century. Out of this work emerged observations and theories that would challenge the views of biologists over the modes of biological evolution over long intervals of geological time. Sudden massive, worldwide extinction events and periods of faunal recovery were documented; these could not be predicted from Darwinian theory. Such events seemed to reset the evolutionary timetable of life in a virtual instant; this prompted some theorists to take a much more 'episodic' or 'contingent' view of the history of life on Earth. Large-scale, or macroevolutionary, changes in global faunal diversity over time seemed to be demonstrable; these again were not predicted from Darwinian theory and required explanation.

Most notably, Niles Eldredge and Stephen Jay Gould proposed the theory of *Punctuated Equilibrium*. They suggested that modern biological versions of evolutionary theory needed to be expanded, or modified, to accommodate patterns of change seen repeatedly among species in the fossil record. These consisted of prolonged periods of stasis (the 'equilibrium' period) during which relatively minor changes in species were observable, and contrasted with very short periods of rapid change (the 'punctuation'). These observations did not fit well with the Darwinian prediction of slow and progressive change in the appearance of species over time (dubbed 'evolutionary gradualism'). These ideas also prompted palaeobiologists to question the levels at which Natural Selection might function: perhaps it could operate above the level of the individual in some instances?

As a consequence the whole field of palaeobiology became more dynamic, questioning, and outward-looking; its practitioners were prepared to integrate their work more broadly with other fields of science. Even highly influential evolutionary biologists such as John Maynard Smith, who had had little truck with fossils at all, were prepared to accept that palaeobiology had valuable contributions to make to the field.

While palaeobiology was re-establishing its scientific credentials, the mid-1960s was also a time of important new dinosaur discoveries, which were destined to spark ideas that are still important today. The epicentre of this renaissance was the Peabody Museum at Yale University (the original workplace of 'bone-fighter' Othniel Charles Marsh). However, this time it was in the person of John Ostrom, a young palaeontologist with a strong interest in dinosaurs.

Chapter 2
Dinosaurs updated

The discovery of 'terrible claw'

In the summer of 1964 John Ostrom was prospecting for
fossils in Cretaceous rocks near Bridger, Montana, and
collected the fragmentary remains of a new and unusual
predatory dinosaur. Further collecting yielded more complete
remains and by 1969 Ostrom was able to describe the new
dinosaur in sufficient detail and to christen it *Deinonychus*
('terrible claw') in recognition of a wickedly hooked, gaff-like
claw on its hind foot.

Deinonychus (Figure 12) was a medium-sized (2–3 metres
in length), predatory dinosaur belonging to a group known
as the Theropoda (see Figures 19 and 29, later in the volume).
Ostrom noted a number of unexpected anatomical features.
These prepared the intellectual ground for a revolution that
would shatter the then rather firmly held view of dinosaurs
as archaic and outmoded creatures that plodded their way
to extinction at the close of Mesozoic world. Unlike others
in his field at that time, Ostrom was more interested in
understanding the palaeobiology of this puzzling animal
than just listing its skeletal features—just as Dollo
might have been (Box 1).

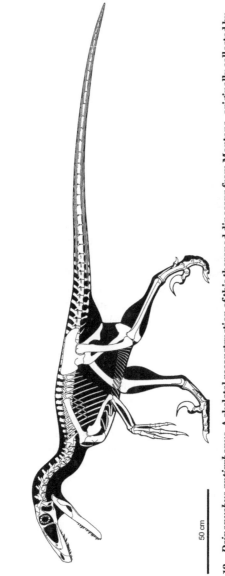

50 cm

12. *Deinonychus antirrhopus*. A skeletal reconstruction of this theropod dinosaur from Montana, originally collected by John Ostrom in the 1960s, displays a range of anatomical features suggestive of a predatory animal with fast movements, good balance, acute senses, and keen intelligence.

Box 1 Features of the *Deinonychus* skeleton

1. The animal was clearly bipedal (it ran on its hind legs alone) and its legs were long and slender, suggesting it was a sprinter.

2. Its feet were unusual in that of the three large toes on each, only two were designed to be walked upon, the inner toe was held clear of the ground and 'cocked' as if ready for predatory action (a bit like a huge version of the sharp retractile claws in a cat's paw).

3. The front part of the animal was counterbalanced at the hip by a long tail; however, this tail was not of the deep, muscular variety normally expected in these types of animal, but was flexible and with muscles present only near the hips, becoming very narrow (and whippy) because it was stiffened by bundles of slender bony rods running along the majority of its length.

4. The chest was short and compact; it supported very long arms that ended in sharply clawed (raptorial) three-fingered hands; these hands swivelled on wrists that allowed the hands to be swung in a raking arc (rather like the arms of a praying mantis).

5. The neck was slender and curved (rather like that of a goose), but it supported a large head that was equipped with long jaws lined with sharp, curved, and saw-edged teeth; very large eye sockets pointed forward (for stereoscopic vision); and the braincase was large, suggestive of unusually high cognitive abilities.

The biology and natural history of *Deinonychus*

The jaws and teeth (sharp, with curved and serrated edges) confirm that this was a predator capable of slicing up and swallowing its prey. The eyes were large, pointed forward, and offered a degree of stereoscopic vision, which would have been ideal for judging distance accurately: very useful for catching fast-moving prey, as well as for monitoring acrobatic movements in three-dimensional space. This also helps to explain its comparatively large brain (implied from its large braincase): the optic lobes would have needed to be large to process lots of complex visual information so that the animal could respond quickly; and the motor-areas of the brain would have needed to be large and elaborate to process the higher brain commands needed to coordinate the rapid muscular responses of the body.

The need for an elaborate brain is further emphasized by considering the light stature and slender proportions of its legs,which are similar to those of modern, fast-moving animals; these suggest that *Deinonychus* was a sprinter. This animal was also quite clearly bipedal, able to walk while balanced on two feet alone (a feat that, as toddlers prove daily, needs to be learned and perfected through feedback between the brain and sensory and musculoskeletal systems). The narrowness of each foot (just two walking toes, rather than the more stable, and more usual, 'tripod' effect of three) further emphasizes that this dinosaur's sense of balance must have been particularly well developed.

Linked to this issue of balance and coordination, the 'terrible claw' on each foot was clearly an offensive weapon: further evidence of the animal's predatory lifestyle. But how, exactly, would it have been used? Two possibilities spring to mind: either it was capable of slashing at its prey with one foot at a time, as large ground-dwelling birds such as ostriches do today (albeit only in defence), which implies that *Deinonychus* could have balanced on one foot from time to time. Or, alternatively, *Deinonychus* may have attacked its prey using a two-footed kick, by jumping on its prey or by grasping its prey in its arms and giving a murderous double-kick—this latter style of fighting is employed by kangaroos when fighting rivals. We are unlikely to be able to establish which of these speculations might be nearest the truth. The long arms and sharp-clawed hands would have been effective grapples for catching, holding, and ripping prey. Also the curious raking motion of the arms, made possible by the unusual wrist joints, enhanced their raptorial abilities. The long, whip-like tail may have served as a cantilever—the equivalent of a tightrope walker's pole to aid balance when slashing with one foot—and it could also have acted as a dynamic *de*-stabilizer, that might have proved very useful when chasing fast-moving, but very elusive, prey.

While this is not an exhaustive analysis of *Deinonychus* as a living creature, it does provide an outline of some of the reasoning that led Ostrom to conclude that *Deinonychus* was an athletic, surprisingly well-coordinated, and rather intelligent predatory dinosaur. However, why should the discovery of this creature be regarded as so important to the field of dinosaur palaeobiology? To answer that question, it is necessary to take a broader view of dinosaurs as a whole.

The traditional view of dinosaurs

Throughout the earlier part of the 20th century, it was widely (and perfectly reasonably) assumed that dinosaurs were just a group of extinct reptiles. Admittedly, some were extraordinarily large or outlandish-looking compared to modern reptiles, but crucially they were still reptiles. Richard Owen (and Georges Cuvier before him) had demonstrated that dinosaurs were anatomically similar to living reptiles, creatures such as lizards and crocodiles. On that basis it was inferred that most of their biological attributes would have been similar to those of these living reptiles: they laid shelled eggs, had scaly skins, and had a 'cold-blooded', or *ectothermic*, physiology.

To help demonstrate that this view was correct, Roy Chapman Andrews had discovered that Mongolian dinosaurs laid shelled eggs, and Louis Dollo (among others) had identified impressions of their scaly skins; so their overall physiology would be expected to resemble that of living reptiles. These attributes created an entirely unexceptional view of dinosaurs: they were scaly, egg-layers, and could be predicted to have been slow-witted and sluggish creatures. Their habits were assumed to be similar to those of lizards, snakes, and crocodiles, which most biologists had only ever seen in zoos (where they indeed seemed slow and sluggish). The only puzzle was that dinosaurs were mostly built on a far grander scale compared to even the very biggest of known crocodiles.

There were many depictions of dinosaurs wallowing in swamps, or squatting as if barely able to support their gargantuan bodies. Some particularly memorable examples, such as O. C. Marsh's *Stegosaurus* and *Brontosaurus*, reinforced these perceptions because both had enormous bodies and the tiniest of brains (even Marsh remarked in disbelief at the 'walnut-sized' brain cavity of his *Stegosaurus*). So lacking in brainpower was *Stegosaurus* that it seemed necessary to invent a 'second' brain in its hip region, to act as a sort of back-up or relay station for information from distant parts of its body: this reinforced the idea of 'stupidity' and the lowly status of dinosaurs.

While all this evidence helped to sustain a popular perception of the dinosaur, contradictory facts were being ignored: some dinosaurs, such as *Compsognathus* (Figure 10) were small, lightly built, and clearly designed for rapid movement: hardly fitting the image of a sluggish and dull-witted animal.

Armed with these conflicting opinions and Ostrom's observations based on his study of *Deinonychus*, it is easier to appreciate how this creature must have been challenging his mind. *Deinonychus* was a large-brained, fast-moving predator capable of sprinting on its hind legs and attacking its prey in extraordinary ways—common sense told him that this was no ordinary crocodile- or lizard-like reptile.

One of Ostrom's graduate students, Robert Bakker, took up this theme by aggressively challenging the view that dinosaurs were dull, stupid creatures. Bakker argued that there was compelling evidence that dinosaurs were in fact similar to today's mammals and birds. It should not be forgotten that this argument echoes the incredibly far-sighted comments made by Richard Owen in 1842, when he first conceived the idea of the dinosaur, and also those of Huxley in the 1860s. Mammals and birds are today regarded as 'special' because they can maintain high activity levels, and this is attributed to their 'warm-blooded' (*endothermic*) physiology. Living endotherms maintain a high and constant body temperature; have highly efficient lungs to maintain sustained aerobic activity levels; are capable of being highly active whatever the ambient temperature; and are able to maintain large and sophisticated brains. All these attributes distinguish birds and mammals from the other vertebrates on Earth.

The range of evidence Bakker used is interesting when considered from our palaeobiological perspective. Using the anatomical observations made by Ostrom, he argued, in agreement with Owen before him, that:

1. Dinosaurs had legs arranged pillar-like beneath the body (as do mammals and birds), rather than legs that sprawl out sideways from the body, as seen in lizards and crocodiles.
2. Dinosaurs could, based on the proportions of their limbs, run at speed using long strides (quite unlike the method used by modern lizards and crocodiles).
3. Some dinosaurs had anatomical evidence of bird-like lungs, which would have permitted them to breathe efficiently—as would be necessary for a highly energetic creature.

However, borrowing from the fields of histology, pathology, and microscopy, Bakker reported that thin-sections of dinosaur bone, when viewed under a microscope, showed evidence of a complex structure and rich blood supply that would have allowed a rapid turnover of vital minerals between bone and blood plasma—such structures strongly resemble those that are seen in modern mammal bones.

Turning to ecology, Bakker analysed the relative abundances of predators and their supposed prey among samples of fossils representing time-averaged communities from the fossil record and the present day. By comparing modern communities of endotherms (cats) and ectotherms (predatory lizards), he estimated that endotherms consume, on average, ten times the volume of prey during the same time interval. When he surveyed ancient (Permian—Figure 3) communities (by counting fossils of this age in museum collections) he observed similar numbers of potential predators and prey. When he examined some dinosaur communities from the Cretaceous Period, he noted that there was a considerably larger number of potential prey compared to the number of predators. He came to a similar conclusion after studying Cenozoic (Figure 3) mammal communities.

Using these simple proxies, he suggested that dinosaurs (or at least those that were predatory) must have had metabolic requirements more similar to those of mammals: for the communities to stay in some degree of balance, there needed to be sufficient prey items to support the appetites of the predators.

Bakker also looked for macroevolutionary evidence (large-scale patterns of change in fossil abundance) in the fossil record. Bakker examined the times of origin and extinction of all dinosaurs (Figure 13) for evidence that might have had a bearing on their putative physiology. The time of origin of the dinosaurs, during the Late Triassic (~230 Ma), coincided with that of a range of quite mammal-like creatures. However, the first mammalian ancestors appeared in ~200 Ma. Bakker suggested that dinosaurs became successful simply because they developed an endothermic metabolism before the first appearance mammals. If not, or so

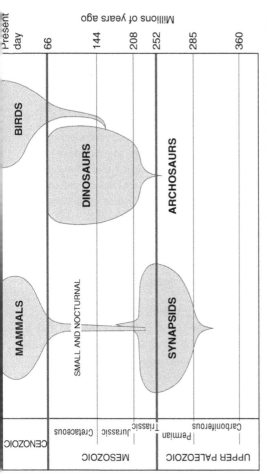

13. The succession of animal life across the Mesozoic. During the Permian and Triassic Periods, *synapsids* were diverse and abundant animals, but toward the end of the Triassic they dramatically reduce in number and variety before giving rise to small, nocturnal, early *mammals*. In marked contrast, the *archosaurs* were not particularly abundant or diverse in Permo-Triassic times, but increased in abundance and diversity toward the end of the Triassic and produced the first *dinosaurs*. The latter evolved spectacularly through the Jurassic and Cretaceous Periods, and also give rise to birds, before spectacularly crashing to extinction 66 Ma, after which the mammals (and dinosaur descendants, the birds) increase in number and diversity through the Cenozoic.

he argued, dinosaurs would never have been able to compete with the first endothermic mammals. In further support of this idea, he noted that early mammals were small, probably nocturnal insectivores and scavengers during the entire Mesozoic Era, when the dinosaurs ruled on land—mammals only diversified into the bewildering variety that we know today once the dinosaurs became extinct at the end of the Cretaceous Period. On that basis dinosaurs simply *had* to be endotherms, otherwise the 'superior' endothermic mammals would have conquered and replaced the dinosaurs in the Early Jurassic. Moreover, when he considered the time of extinction of the dinosaurs at the close of the Cretaceous (66 Ma), Bakker believed that there was evidence that the world had been subjected to a temporary period of low global temperatures. Since dinosaurs were, in his opinion, large, endothermic, and 'naked' (that is, they were scale-covered and had neither hair nor feathers to keep their bodies warm), he reasoned that they were unable to survive a period of rapid climatic cooling and therefore died out. This left the mammals and birds to survive to the present day. Dinosaurs were too big to shelter in burrows, as do the small reptiles that evidently did survive the end-Cretaceous changes.

Bakker therefore proposed that far from being slow and dull, dinosaurs were intelligent, highly active creatures that had 'stolen' the world from the mammals for the remaining ~135 million years of the Mesozoic. Rather than being ousted from the world by the evolutionary rise of 'superior' mammals, they had only given up their dominance because of some freakish worldwide cooling event sixty-six million years ago.

Clearly the palaeobiological approach to research has suddenly become intellectually broad-based. The 'expert' can no longer rely upon one specialist area of knowledge—his or her expertise must encompass anatomy, comparative biology, physiology, histology, microscopy, ecology, and stratigraphic geology...just for starters!

Ostrom and *Archaeopteryx*: the earliest bird

Having described *Deinonychus*, Ostrom continued to investigate the biological properties of dinosaurs. In the early 1970s a trifling discovery in a museum in Germany was to bring him right back to the centre of some heated discussions. While examining collections of flying reptiles, Ostrom noticed one specimen, collected from a quarry in Bavaria that did not belong to a pterosaur (flying reptile), as its label suggested. It was a section of a leg including the thigh, knee-joint, and shin. Its detailed anatomical shape reminded Ostrom of the leg of *Deinonychus*. On closer inspection, he could also make out the faintest impressions of feathers. This was clearly an unrecognized specimen of the fabled early bird *Archaeopteryx* (Figure 9). Excited by his new discovery, and naturally puzzled by the apparent similarity to *Deinonychus*, Ostrom restudied all the known *Archaeopteryx* specimens.

The more Ostrom studied *Archaeopteryx*, the more convinced he became of the similarity between this creature and *Deinonychus* (Figure 12). This led him to reassess the authoritative work on bird origins that had been written by ornithologist and anatomist Gerhard Heilmann in 1927. The sheer number of anatomical similarities between carnivorous theropod dinosaurs and early birds drove Ostrom to question Heilmann's view that any similarities were the result of evolutionary convergence: that is to say that their bodies looked similar, but they were not related to one another.

Encouraged by his discoveries and more detailed observations of theropods and *Archaeopteryx*, Ostrom published a series of articles in the 1970s that led to the acceptance of a theropod dinosaur ancestry of birds by the majority of palaeontologists. Ostrom's conclusions would have delighted the far-sighted Huxley and (probably) have infuriated Owen.

The close anatomical, and by implication *biological*, similarity between some theropods and the earliest birds added more fuel to the controversy concerning the metabolic status of dinosaurs. Birds are known to be very active, feathered, endothermic creatures. If birds evolved from theropod dinosaurs, where exactly does one draw a line between a creature being identified as theropod or bird? The once clear dividing line between feathered birds, with their distinctive anatomy and biology that merited them being separated off from all other vertebrates as the Class *Aves*, discrete from the Class *Reptilia* (of which the dinosaurs were just one extinct group) was becoming potentially blurred (see Chapter 5).

Chapter 3
A new perspective on *Iguanodon*

The resurgence of palaeobiology in the 1960s, and the new insights into dinosaurs prompted by John Ostrom's important work, acted as a spur to reinvestigate some of the earliest discoveries.

Louis Dollo's description of the incredible discoveries of *Iguanodon* at Bernissart had created the image of a giant (5 metres tall; 11 metres long) kangaroo-shaped creature (Figure 9). It had:

> powerful back legs and a massive tail that helped it to balance...[and] was a plant eater...it grasped bunches of leaves with its long tongue, then pulled them into its mouth to be clipped off with the beak.

Iguanodon was the dinosaur equivalent of a 'tree browser', represented in the recent past by the giant extinct South American ground sloths or today's giraffes. Dollo himself referred to *Iguanodon* as a 'girafe reptilienne'. Rather surprisingly, nearly every aspect of Dollo's vision of *Iguanodon* is either incorrect or seriously misleading.

Bernissart: a ravine where *Iguanodon* perished?

Some of the earliest work at Bernissart focused on the extraordinary circumstances of the original discovery. The dinosaurs had been unearthed in a coal mine at depths of between 322 and 356 metres below the surface. This was unexpected, as the coal seams being excavated were known to be of Paleozoic age (Figure 3) yet dinosaurs are unknown in rocks of such antiquity. However, the *Iguanodon* skeletons were not found in the ancient coal seams, but in laminated clays of Cretaceous age that cut across the more ancient coal-bearing rocks. Mining geologists had a commercial interest in discovering the extent of these clays, and the degree to which they might affect coal extraction, so they began mapping the area.

Cross-sections of the mine, created during these geological investigations, showed that laminated clay cut through the Paleozoic rocks (with their valuable coal seams). The cross-sections showed steep-sided 'ravines' that seemingly cut into the ancient rocks. This led to the rather appealing notion that the Bernissart dinosaurs represented a herd that had tumbled to their death. Dollo, himself no geologist, was more inclined to the idea that these dinosaurs had lived, and died, in a narrow gorge. However, the more dramatic story had the greater impact, and was further embellished by suggestions that they had been stampeded into the ravine by huge predatory dinosaurs or by some freak event such as a forest fire. This was not entirely wishful thinking: extremely rare fragments of a large predatory dinosaur were discovered within the *Iguanodon*-bearing beds; and charcoal-like lumps of coal were recovered from some of the rubbly deposits found in the region between the coal-bearing rock and the dinosaur-bearing beds. So, what really happened?

The discoveries at Bernissart presented a huge logistic challenge in the 1870s and early 1880s. Complete skeletons of dinosaurs measuring up to 11 metres in length had been discovered at the bottom of this deep mine; they were the focus of worldwide interest at the time, but how were they to be excavated successfully and then studied?

A co-operative venture was established between the Belgian government (funding the scientists and technicians of the Royal Natural History Museum in Brussels), and the miners and engineers at the colliery in Bernissart. Each skeleton was carefully exposed and its position in the mine recorded systematically on plan diagrams. Every skeleton was divided into movable blocks (approximately 1 metre square). Each block, protected by a jacket of plaster of Paris, was carefully numbered and recorded on plan drawings (Figure 14(a)) before being lifted and transported to Brussels.

In Brussels, the blocks were reassembled, like gigantic jigsaw pieces, and the plaster painstakingly removed to reveal the bones of each skeleton. At this point an artist drew the bones before any further preparation or extraction was undertaken (Figure 14(b)). Some skeletons were completely extracted from the clay and mounted to create a magnificent display that can be seen today at the Royal Belgian Institute of Natural Sciences in Brussels. Other skeletons were cleared of the clay on one side only and arranged in their burial position on wooden scaffolding supporting vast banks of plaster. This display mimics their entombed positions when they were first discovered in the mine at Bernissart.

The original plans of each excavation, as well as geological sections are preserved in the archives in Brussels. This information was 'mined' by the author for evidence concerning the geological nature of the dinosaur burial site.

The geology of the Mons Basin, which includes the village of Bernissart, had been studied before dinosaurs were ever discovered. In 1870 it was known that the coal-bearing strata were pock-marked by 'cran' (natural subterranean sink-holes). Each 'cran' was roughly circular and filled with clay. It was concluded that these had formed by dissolution of Paleozoic rocks deep underground, creating large caverns. The roofs of such caverns collapsed periodically under the sheer weight of the

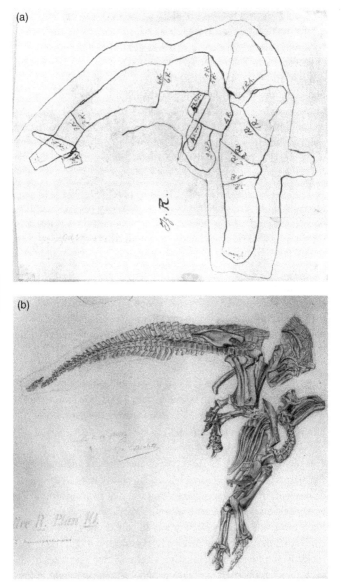

overlying rocks, so the caverns became filled with whatever lay above: in this case pliable shaley clay. Collapses had been reported in the Mons area as earthquake-like vibrations. By coincidence a minor earthquake of this type took place during August 1878 while the dinosaurs were being excavated. Minor collapses were seen in the tunnels and some flooding occurred, but no miners or scientists perished and they were able to resume work once the flood water had been pumped away.

Despite all this local geological knowledge, it is strange to discover that the scientists from the museum in Brussels incorrectly interpreted the geological nature of the 'cran' at Bernissart. The mining engineers produced geological sections from the tunnels that yielded the dinosaurs; these showed that immediately beyond the coal-bearing seams there was a region of 10–11 metres of breccia (broken beds containing irregular blocks of limestone and coal mixed with clay) before entering steeply dipping, but more regularly layered, clays containing the fossils. Toward the middle of the 'cran' the clay beds became horizontally bedded, and as the tunnel approached the opposite side of the 'cran' the beds once again became steeply tilted in the opposite direction before passing again into another brecciated region and finally re-entering the coal-bearing limestone. The symmetry of the geology across the 'cran' is exactly what would be expected if overlying sediments had progressively slumped into an underlying cavern.

14. *Iguanodon* (a) **This is an original plan diagram that was made by the team of miners and scientists as they excavated dinosaur skeletons in the mine at Bernissart. The shape of the blocks and letter-codes enabled the blocks to be reassembled accurately after their transportation to the workshop/laboratory in Brussels. (b) In Brussels the plan diagram (Figure 14(a)) allowed the blocks to be reassembled. Each block was carefully prepared so that the fossil bones were made visible. The artist Gustave Lavalette then prepared an accurate sketch before any further extraction was undertaken. This sketch reveals an entire dinosaur skeleton.**

The sediments in which the dinosaurs are embedded also directly contradict the ravine or river-valley interpretations. Finely laminated clays containing the fossils are normally deposited in low-energy environments, typically large lakes or lagoons. There is no evidence for catastrophic deaths caused by herds of animals plunging into a ravine. In fact the dinosaur skeletons were found in separate layers of sediment (along with fish, crocodiles, turtles, thousands of leaf impressions, and even rare insect fragments), proving that they definitely did not all die at the same time and could never have been part of a single catastrophic event.

Study of the orientation of the fossil skeletons within the mine revealed that dinosaur carcasses were washed into the burial area on separate occasions and from different directions. It was as if the direction of flow of the river, which carried the dinosaur carcasses and other remains, had changed from time to time, exactly as happens in large, slow-moving river systems today.

So, as early as the 1870s, it was clearly understood that there were neither 'ravines' nor 'river valleys' in which the dinosaurs at Bernissart might have perished. It is fascinating how the 'dramatic' discovery of dinosaurs at Bernissart seems to have demanded an equally dramatic explanation for their deaths, and that such fantasies were uncritically adopted even though they flew in the face of the scientific evidence available at the time.

The image of *Iguanodon* as a gigantic kangaroo-style creature has become iconic because of the generous distribution of full-sized skeletal casts to many museums around the world. But does the evidence for this posture survive further scrutiny?

A 'twist' in its tail

By re-examining the skeletons of *Iguanodon* I was able to reveal some disconcerting features. One of the most obvious of these concerns the massive tail of *Iguanodon*. This well-known ('iconic') reconstruction shows the animal (Figure 8) propped, kangaroo-style, on its tail and hind legs. In this posture, the tail curves upward to the hip. In sharp contrast, all the documentary and fossil evidence points to this animal holding its tail essentially straight or somewhat downwardly curved. This is clearly seen in the specimens arranged on banks of plaster in the museum, and in the wonderful pencil sketches made of their skeletons before they were exhibited (Figure 14(b)). It could of course be argued that this shape was simply an artefact of preservation, but this explanation is definitely not plausible here. The backbone was in effect 'trussed' on either side by a trellis-like arrangement of long bony tendons that tensioned and held the backbone nearly straight; some of these bony tendons can be seen in Figure 14(b). As a result, the heavy, muscled tail acted as cantilever to balance the weight of the front part of the body at the hips. The truth is that the upward sweep of the tail seen in Dollo's reconstructions would have been physically impossible for these animals in life. The tail was deliberately broken in several places to achieve the upward bend—a case perhaps of Louis Dollo making the skeleton fit his personal ideas a little over-zealously!

This observation perturbs the pose of the remainder of the skeleton. If the tail is straightened so that it can adopt a more natural shape, then the tilt of the body changes dramatically: the backbone becomes more horizontal and balanced at the hip. As a result the chest is lowered, bringing the arms and hands closer to the ground.

Hands as feet?

The hand of *Iguanodon* is part of dinosaurian folklore for one obvious reason. The conical thumb-spike was originally identified as a rhinoceros-like horn on the nose of the *Iguanodon* (Figure 6) and immortalized in the giant concrete models erected at London's Crystal Palace (Figure 1). Dollo showed that this spiky 'horn' was the dinosaur's thumb. However, the hand (and the entire forelimb) of this dinosaur held a few more surprises.

The thumb-spike sticks out at right-angles to the rest of the hand and can be moved very little (Figure 15). The second, third, and fourth fingers are very differently arranged: three long bones (metacarpals) form the palm of the hand and are bound tightly together by strong ligaments; the fingers are jointed to the ends of these metacarpals and are short, stubby, and end in flattened, blunt hooves. When these bones were manipulated to see what their true range of movement was likely to be, I found that the fingers splayed outwards (away from each other) and certainly could not flex to form a fist or perform a grasping structure, as might have been expected. This arrangement resembles that seen in the *feet* of this animal: the three central toes of each foot are similarly shaped and jointed, splay apart, and end in flattened hooves. The fifth finger (Figure 15) is different from all the others: it is quite separate from the previous four and set at a wide angle from the remainder of the hand; it is also long and has a wide range of movement at each joint, and was presumably extremely flexible.

This re-examination led me to dramatically revise earlier ideas and conclude that the hand is one of the most peculiar seen in the entire animal kingdom. The thumb was without doubt an impressive, stiletto-like weapon of defence; the three central fingers were clearly adapted to bear weight—rather than for grasping as hands usually do—while the fifth finger was sufficiently long and flexible to be used for grasping (Figure 15).

The idea that the hand could act as a foot for walking upon, or at least supporting some of the body weight, was revolutionary—but was it true? This prompted further research on the arm and shoulder for additional evidence that might confirm such a radical reinterpretation.

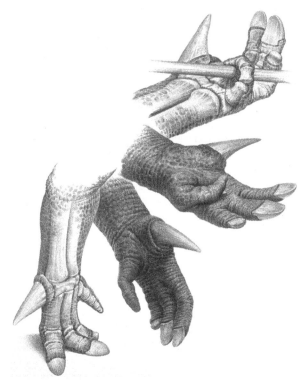

15. *Iguanodon*. **The functions of the hand summarized. The inner finger (thumb) was a large conical spike, the central three fingers were splayed apart and ended in flattened hooves (for walking upon), and the fifth finger was long, slender, and prehensile for grasping.**

The wrist showed that its bones are welded together to form a solid block, instead of being smooth, rounded bones that could slide past one another in order to allow the hand to swivel against the forearm. The wrist bones were locked together by a sort of bony cement and were further strengthened by strands of bony ligament. These features obviously combined to lock the wrist against the hand and forearm bones to resist the forces acting through them during weight-bearing.

The arm bones are extremely stoutly built, again primarily for strength during weight support, rather than for allowing flexibility as is more normal with genuine arms. The upper arm (humerus) is massive and had huge arm and shoulder muscles. The stiffness of the forearm has important consequences for the way in which the hand would have been placed on the ground—the fingers would have pointed outward and the palms inward—an unusual consequence of converting a hand into a foot (Figure 16). The pose of the hand has been confirmed by examination of the shape of forefoot prints left by this dinosaur. The arm is unusually long, over three-quarters the length of the hind limb. The true size of the arms was somewhat masked in the original skeletal reconstructions because they were folded against the chest and always *seemed* shorter than they really are.

Finally, the shoulder bones are large and powerfully built and show another unexpected feature. In the centre of the chest of the larger skeletons at Bernissart there is an irregular bone that grew in the soft tissues across the centre of the chest. This bone (that I named an *intersternal ossification*) is pathological in origin—it is very irregular and formed in response to strain in the soft tissues of the chest while the animal was walking on all fours.

Reassessing the posture of *Iguanodon* in the light of these observations, it seems clear that a more natural pose of the backbone was horizontal, with the body weight distributed along the backbone and largely balanced at the hips and supported by the massive and strong hind legs (Figure 16). The ossified tendons distributed along the spine, above the chest, hip, and tail, and acted as tensioners to distribute body weight along the backbone. This pose allowed the front limbs to reach the ground for use in weight support while these animals were stationary. *Iguanodon* probably moved slowly on 'all-fours' at least part of the time (Figure 16).

Careful study of the abundant large *Iguanodon* skeletons from Bernissart revealed that a few were smaller than the average.

16. *Iguanodon*. A restoration of an adult *Iguanodon bernissartensis* walking quadrupedally.

Measuring the proportions of each of these skeletons revealed an unexpected growth change: smaller, presumably immature, specimens had shorter arms than would have been expected. The comparatively short-armed juveniles may well have been more adept bipedal runners, but as large adult size and stature was achieved they became progressively more accustomed to moving on all-fours.

Iguanodon had a variety of close relatives

The Bernissart discoveries are notable for comprising two varieties of *Iguanodon*-like dinosaur. One (*Iguanodon bernissartensis*—'*Iguanodon* from Bernissart') is large and robustly built (see Figure 16) and represented by more than thirty-five skeletons; the other (*Mantellisaurus atherfieldensis*—'Mantell's reptile from Atherfield') is smaller and more delicately built (approximately 6 metres in length) and represented by only two skeletons. In recent years it has come to light that two other *Iguanodon*-like (iguanodontian) dinosaurs lived in Sussex a few million years before *Iguanodon* and *Mantellisaurus*. I named these early species *Barilium* ('heavy hip bone') and *Hypselospinus* ('high spine') and as more discoveries are made it seems clear that there was a substantial diversity of iguanodontians living in various parts of the world during Early Cretaceous times.

Soft tissues

Soft tissues of fossil creatures are preserved only rarely and under exceptional preservational conditions. Louis Dollo reported small patches of skin impressed on parts of the skeletons at Bernissart. The texture of the skin is of a finely scaled, flexible covering very similar in appearance to that seen on the skin of modern lizards.

Brain tissues

Some broken fossils collected from an intertidal pool on the coast of Sussex, although not well preserved, belonged to a small iguanodontian dinosaur (probably either *Barilium* or *Hypselospinus*). Among the remains was a small pebble with a curious shape. The 'pebble' proved to be an extraordinarily lucky find: it was a partial internal cast (endocast) of the braincase of a dinosaur. Parts of the surface of the endocast examined with a hand-lens had textures that seemed worth detailed examination using a scanning electron microscope (SEM) and a MicroCT (micro computed tomography) scanner. Energy Dispersive X-ray Spectroscopy (EDS) was also used to assess the elements and mineral composition of the surface features. What emerged were remarkable details of the tough membranes (meninges) that surround the brain, the blood vessels that ran through these meninges, and a hint of the superficial parts of the cortex of the brain and its associated capillaries.

These soft-tissue structures were preserved because the brain tissues had been 'pickled' during the process of death, decay, and burial of this unfortunate dinosaur. These hardened membranes, blood vessels, and soft tissues were gradually replaced by microscopic crystals of calcium phosphate (collophane) and iron carbonate (siderite). Analysis of the chemistry associated with normal decay processes allowed us to appreciate that this animal had perished and its body had fallen into stagnant water; the combination of stagnant conditions and the chemicals released during decay created an environment in which such extraordinary preservation was able to take place. In many ways this sort of analysis bears remarkable similarity to that used in modern forensic science.

A complete brain cavity cast of *Iguanodon*

While the extraordinary 'pebble' just described is an incomplete portion of the entire brain cavity of an iguanodontian with a few traces of the superficial membranes preserved as mineral coatings, other specimens give a more complete picture of the shape of the entire brain cavity. An *Iguanodon* specimen in the collections of the Natural History Museum in London comprises a pretty much complete cast of the brain cavity of this dinosaur. The original specimen was a very battered partial skull of the dinosaur. However after the animal had died, and its soft tissues had decayed the muddy sediment in which the skull was buried filled all the internal spaces of the skull and hardened (lithified) to a concrete-like consistency. The clay matrix in which the skull was buried was impermeable, so ground-water containing minerals was unable to seep through the clay and petrify the skull bones, so they were comparatively soft and crumbly. I was able to carefully remove the crumbly skull bone and expose the shape of the internal spaces in the skull as a natural cast; these included the cavity where the brain had lain, the space for the inner ear, and many of the passages occupied in life by blood vessels and nerves that led to and from the brain cavity.

Iguanodon's diet

The first recognizable fossils of *Iguanodon* were teeth, whose features showed that it was a herbivorous animal: they were chisel-shaped to be able to slice plants in the mouth before they were swallowed.

Carnivores eat meat. From a biochemical and nutritional perspective meat is one of the simplest and most obvious options for any creature. Most of the other creatures in the world are made of roughly similar chemicals as the carnivores that eat them. Their flesh is therefore a ready and rapidly assimilated source of food, provided the prey can be caught, sliced into chunks (or even swallowed whole), and then quickly digested in the stomach. This whole process has the potential to be relatively quick and efficient.

Herbivores face a more challenging situation. Plants are neither particularly nutritious nor readily digested compared to animal flesh; they are built of cellulose, a material that gives them strength and rigidity. The inconvenient point about this unique chemical is that it is *completely indigestible*: there is nothing in the armoury of enzymes in our guts that can dissolve cellulose. As a result, the cellulose portion of plants passes straight through animals' guts as what we call roughage. So, how do herbivores survive on what appears to be such an unpromising diet?

Plant-eaters have successfully adapted to this diet by exhibiting a number of characteristic features. They have teeth with hard-wearing grinding surfaces and powerful jaws to grind plant tissues. Chewing releases the digestible contents of each cellulose-walled plant cell. Herbivores eat large quantities of plant food in order to be able to extract sufficient nutrients and as a consequence have barrel-shaped bodies. Herbivore guts are also modified to provide spaces (chambers in which microbes can live); our appendix is a tiny vestige of such a chamber, and hints at a plant diet in our primate ancestry. These *symbiotic* microbes are able to synthesize *cellulase*, an enzyme that converts cellulose into simple sugars that can then be absorbed by the host animal.

Iguanodon (11 metres long and weighing ~7 tonnes) was a herbivore that consumed large quantities of plants. But how, precisely, did *Iguanodon* process this food?

Iguanodon chewed its food

At the front of its jaws was a sharp horny beak supported by the *predentary* and toothless part of the upper jaw (Figures 17 and 18). Behind, the long jaws are lined with parallel rows of chisel-shaped teeth that form cutting blades (Figure 17). Each tooth rests neatly against its neighbours in a rank-and-file arrangement, and beneath the working teeth are replacements that will slot into place as the teeth above get worn away. This continuous replacement pattern is normal for reptiles in general. What is unusual, even by reptile standards, is that the teeth are held together in an ever-growing elongate blade. Wear between opposing (upper and lower) blades maintains a cutting surface throughout the life of the dinosaur. Rather than having permanent, hard-wearing teeth (as we do), this could be described as a disposable model that relies on constant replacement of individually simpler teeth.

Opposing edges of each cutting blade of teeth have characteristics that ensure efficiency in their cutting action. The inner surfaces of the lower teeth are coated in a thick layer of extremely hard enamel, while the remainder of the tooth is made of softer dentine and a pulp cavity packed with softer bony material. In contrast, the upper teeth have the reverse arrangement: the *outer* edge is coated in thick enamel and the remainder of the tooth is composed of dentine and bony infill. When the jaws close the enamelled leading edge of each lower teeth meets that of the upper teeth rather like the cutting edges of a pair of scissors (Figure 17).

Once the enamelled edges have passed one another, the enamel edges (unlike scissor blades) then cut against the slightly softer dentine and bone of the remainder of the tooth in a tearing and grinding movement that crushes plant tissues.

The geometry of the cutting surfaces of the upper and lower teeth is interesting. The worn surfaces are oblique (Figure 18(c), (d)), the lower surfaces face outward and upward, while the upper teeth have worn surfaces that face inward and downward. In living reptiles, the closure of the lower jaw is brought about using a simple hinge: the jaws on either side of the mouth closing simultaneously in what is called an *isognathic* bite driven by muscles attached to the *coronoid process* (Figure 18). If this type of bite is proposed for *Iguanodon*, then it is obvious that the two sets of teeth on either side of the mouth would become immovably wedged together: the lower jaws would simple jam tightly against the upper ones: this has to be wrong!

For angled wear surfaces to develop on the teeth there would have had to be some sideways movement by the jaws as they closed.

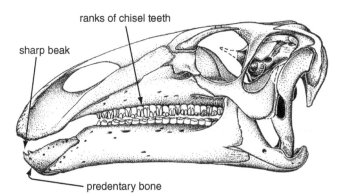

ranks of chisel teeth

sharp beak

predentary bone

17. *Iguanodon*. A reconstruction of the skull of *Iguanodon* showing the curious toothless beak at the tip of the jaws, the long rows of interlocking chewing teeth, as well as the horizontal, trough-like recess on the side of the mouth where a fleshy cheek would have existed in life.

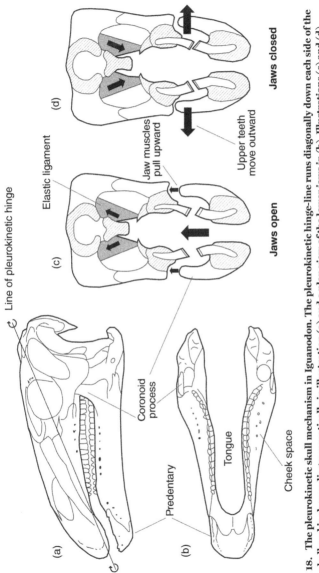

Line of pleurokinetic hinge

Coronoid process

Predentary

Tongue

Cheek space

Elastic ligament

Jaw muscles pull upward

Upper teeth move outward

Jaws open

(c)

Jaws closed

(d)

(a)

(b)

18. The pleurokinetic skull mechanism in Iguanodon. The pleurokinetic hinge-line runs diagonally down each side of the skull and is shown diagrammatically in illustration (a) and a plan view of the lower jaws in (b). Illustrations (c) and (d) represent cross-sections of the dinosaur's head.

Just this type of movement is seen in living herbivorous mammals, they have an *anisognathic* jaw closure mechanism: their lower jaws are closer together than the upper jaws and special muscles, arranged in a sling on either side of each jaw, control the position of the jaws very precisely. As a result the teeth on one side meet one another and the lower set is forcibly slid inwards so that the teeth grind against one another, then the jaw is repositioned and the teeth on the opposite side do the same: this is chewing! We humans employ this type of jaw motion, especially when eating tough foods, but it is far more obvious in some classically herbivorous mammals (e.g. cows, sheep, and goats), in which the lateral swing of the jaw is very pronounced.

The mammalian chewing mechanism depends upon complex jaw muscles and its nervous control system, as well as specially reinforced skulls capable of withstanding the stresses created by this chewing method. In contrast *Iguanodon* did not possess an anisognathic jaw arrangement, it lacked the complex muscular arrangement that allows the lower jaw to be very precisely positioned (whether they had the nervous system to control such movements is largely irrelevant), and its skull is not specially reinforced to cope with the induced skull stresses.

Iguanodon is a puzzle: it does not conform to the expected models of jaw movement. Is its anatomy wrongly interpreted? Or was this dinosaur doing something unusual?

Solving the puzzle

The lower jaws of *Iguanodon* are strong, and the front end is clamped to its neighbour by the predentary bone (Figure 18). The teeth run along the length of the jaw, and at the rear there is a tall prong (coronoid process) that acts as an attachment area for powerful jaw-closing muscles and as a lever that increases the force of jaw closure. During biting, the upper jaws would have been subjected to vertical forces, created by the closure of the lower jaws against the uppers, as well as *sideways* forces generated by the lower teeth wedging themselves between the upper teeth (Figure 18(c) and (d)).

Of all forces acting on the skull of *Iguanodon*, the ones that it is least well-equipped to deal with are *sideways* forces acting on its teeth. The long snout resembles a deeply vaulted roof. To resist sideways forces acting on the teeth, the skull would need to be braced by bony 'joists' connecting the upper jaws; this is the arrangement (called the secondary palate) found in living mammals. Without such bracing, the skull of *Iguanodon* is very vulnerable to splitting along its midline simply because the cheek-bones act as long levers against the roof of the snout. Midline breakage of the skull was avoided by the provision of hinges that are arranged diagonally along either side of the skull (Figure 18(a)); these allow the sides of the skull to flex outward as the lower teeth force their way between the uppers (Figure 18(d)). Other features deeper within the skull helped to provide control over the amount of movement that was possible along this hinge (so that the upper jaws did not simply flap around loosely).

I named this remarkable mechanism *pleurokinesis* ('side motion'). The pleurokinetic mechanism permits a *grinding* motion between opposing sets of teeth: the upper teeth swinging outward across the lowers, which mimics the chewing motion achieved by herbivorous mammals, but in a completely different way.

This novel chewing system explained another important observation: *Iguanodon*'s teeth are recessed (set inwards—Figures 17 and 18) from the side of the face. This creates a trough that would have been enclosed by a fleshy cheek—another most un-reptilian feature. Given that the upper teeth slid past the lowers to cut up their food, it seems logical to expect that every time they bit through food in the mouth, at least half of it would be lost from the sides of the mouth...unless, of course, it was caught by a cheek and then pushed back in. So these dinosaurs appeared to be not only capable of chewing their food in a surprisingly sophisticated way, but they also had mammal-like cheeks, and of course to aid the positioning of the food between the teeth between chews, they would have needed a long, muscular tongue.

Once this new chewing system had been identified, I was able to recognize that *pleurokinesis* was not a 'one-off' invention by *Iguanodon*. It was actually widespread among the group of dinosaurs known as ornithopods, to which *Iguanodon* belonged. Tracing the general evolutionary history of ornithopods across the Mesozoic Era, it became clear that these types of dinosaur became increasingly diverse and abundant in time. Ornithopods (known as hadrosaurs—see Figures 19 and 29, later in the volume) reached their greatest abundance and diversity in the latest Cretaceous Period. Some bone-bed discoveries in North America hint at herds of hadrosaurs numbered in the thousands. Hadrosaurs had incredible grinding teeth (each jaw contained up to 1,000 teeth) and a well-developed pleurokinetic mechanism.

It seems probable that one of the factors that allowed hadrosaurs to become abundant and diverse was their feeding efficiency. Their evolutionary success was in part at least a consequence of their inheritance of the pleurokinetic chewing mechanism identified among their ancestors such as *Iguanodon*.

Chapter 4
Relationships between dinosaurs

One of the important tasks undertaken by palaeontologists is to try to discover the genealogy, or evolutionary history ('tree of life', if you will) of the species that they study. To put dinosaurs as a whole into some sort of perspective, it will be necessary to outline the techniques used to do this, and our current understanding of dinosaurian evolutionary history.

One feature of the fossil record is that it offers the tantalizing possibility of tracing the genealogy of organisms not just over a few human generations (as modern genealogists might) but over thousands, or millions, of generations, across the immensity of geological time. The primary means by which such research is carried out at present is the technique known as *cladistics*. The premise of this technique is really quite simple. It accepts that organisms are subject to the general processes of Darwinian evolution. This does not require anything more profound than the assumption that organisms that are more closely related, in a genealogical sense, tend to physically resemble each other more closely than they do more distantly related creatures. To try to investigate the degree of relatedness of creatures (in this particular case, fossil creatures), *palaeosystematists* are most interested in identifying as wide a range of anatomical features as are preserved in the hard parts of their fossils. Unfortunately, a great deal of really important biological information has simply decayed and been lost during the process of fossilization of any skeleton, so, being pragmatic about things, we simply have to make the most of what is left. Until quite recently, the reconstruction of phylogenies has relied on hard-part anatomical features of animals alone; however, technological innovations have now made it possible to compile data, based on the biochemical and molecular structure of living relatives, that can add significant and new information to such investigations.

What the dinosaur systematist has to do is compile detailed lists of anatomical characteristics, with the intention of identifying those that are phylogenetically important; that is to say they contain an evolutionary signal. The task is intended to produce a workable hierarchy of relationship, based on groupings (*clades*) of ever more closely related animals.

The analysis also identifies features that are unique to a particular fossil species; these are important because they establish the special characteristics that, for example, distinguish *Iguanodon* from all other dinosaurs. This probably sounds blindingly obvious but, in truth, fossil creatures are often based on a small number of bones or teeth. If other partial remains are discovered in rocks elsewhere from the original, but of very similar age, it can be quite a challenge to prove convincingly whether the new remains belong to, say, *Iguanodon*, or perhaps to a new and previously undiscovered creature.

Beyond the features that identify *Iguanodon* as unique, there is also a need to identify anatomical features that it shares with other equally distinct, but quite closely related, animals (other members of the same clade). You might say that these were the equivalent of its anatomical 'family'. The more general the characters that 'family' groups of dinosaurs share, the more this allows them to be grouped into ever larger and more inclusive categories of dinosaurs that gradually piece together an overall pattern of relationships for them all.

The real question is: how is a *cladogram* or pattern of relationships created? The technique is known as *cladistics* (phylogenetic systematics) and is capable of generating branching (tree-like) diagrams of relationship known as cladograms (e.g. Figure 19). It must be emphasized that these are *hypothetical* patterns of relationship: they do not necessarily represent 'the truth' they are simply an attempt to probe the genealogy of (in this case dinosaurs), but once produced are then open to scientific criticism, scrutiny, and discussion.

Figure 19 is a cladogram of dinosaur relationships; it has a branching tree-like structure that links together the principal groups of dinosaurs mentioned in this book (see also Figure 29, later in the volume). To create this, researchers need to compile a table (*data matrix*) containing a column listing the species under consideration and against this a compilation of the features/characters (anatomical, primarily) that each species exhibits. Each species is 'scored' in relation to whether it does (1) or does not (0) possess each character, or in some instances if the decision is uncertain this can be signified as a (?) or a (–) if it is entirely absent. The resulting matrix of data (these can be large) is then analysed using a variety of proprietary computer programs, whose role is to assess the distribution of 1s and 0s and generate a set of statistics that determine the most parsimonious distribution of the data that are shared by the chosen species. The resulting cladogram (or cladograms) form(s) the starting point for further investigation that is aimed at revealing common patterns or overall similarities, and the extent to which the data might be misleading or erroneous. The cladogram that results from this type of analysis represents no more than a *working hypothesis* of the relationships of the animals that are being investigated. One approach that attempts to find areas of agreement among independently generated cladograms is to compile all the data from these cladograms and search for areas where they seem to agree: this is known as the *supertree* analysis.

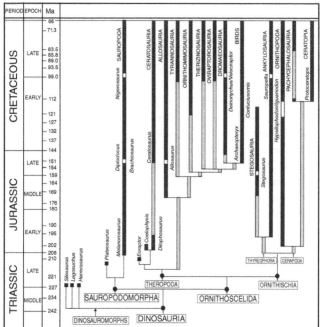

19. Dinosaur cladogram, calibrated against the geological timescale in order to indicate the approximate time of origin of the various clades as well as their known duration within the fossil record.

Each branch on the tree marks a point at which it is possible to define a group of species (the clade) that are connected by sharing a number of characteristic features. Using this information it is possible to construct what is, in effect, a type of genealogy (*phylogeny*) that represents a model of the evolutionary history of the group as a whole. If the known geological times of occurrence of each of the species are plotted onto this pattern, it becomes possible to indicate the overall history of the group and suggests the probable time at which various of the species may have originated (Figure 19). In this way the cladogram, rather than simply representing a statistically supported arrangement of species, begins to resemble more strongly a genealogy. Obviously, each phylogeny created in this way is only as good as the data available, but research moves on constantly and the data and how it is scored and analysed changes: following the discovery of new, better, or more complete fossils; depending on how the data is scored by researchers; and also with the advent of new or improved methods of analysis.

Figure 19 represents the conclusion of a very recent re-analysis of the genealogy of dinosaurs that reflects new work by my research student Matthew Baron, former student Paul Barrett and myself in Cambridge. These ideas revolutionize views that have remained unchallenged since the work of another Cambridge-trained palaeontologist (Harry Govier Seeley) in 1887. And if these very new ideas are agreed and adopted it means that all textbooks about dinosaurs are going to have to change!

An evolutionary history of the dinosaurs: in brief

Cladistic analysis of the relationships between dinosaurs was first initiated by Jacques Gauthier (Yale University) in 1984 based upon an analysis of a range of archosaurs (i.e. dinosaurs and their near relatives). I also published a cladogram focused on ornithischians in 1984 and its topology (the general shape of that tree) has been supported and refined subsequently.

The clade *Dinosauria* is traditionally recognized (as Owen so perceptively realized) as a clade of reptiles with upright leg posture and specially reinforced connections between the hips and vertebral column to ensure the efficient carriage of the body on its legs. These changes conferred upon early dinosaurs some valuable assets: pillar-like legs could support great body weight, so dinosaurs could become very large creatures; and, upright legs also permit a long stride, meaning that some dinosaurs had the potential to move very quickly. Both these attributes were used by different dinosaur sub-clades throughout their reign on Earth.

While dinosaurs share these crucial anatomical features, they have, since the work of Seeley in 1887, been separated into two fundamentally distinct sub-clades: *Saurischia* ('lizard-hipped') and *Ornithischia* ('bird-hipped'). As these names suggest, the differences are to be found in the arrangement of the bones forming their hips (Figure 20). Saurischians included two sub-divisions: *Theropoda* (essentially bipedal carnivores) and *Sauropodomorpha* (mostly quadrupedal herbivores). *Ornithischians* were an extremely morphologically diverse group of herbivores. The earliest members of these groupings of dinosaur have been identified in rocks of Norian age (about 220 Ma). However in recent years a steadily increasing number of slightly more ancient fossil reptiles have been discovered and studied that seem to be 'not-quite-dinosaurs' and are referred to as *dinosauromorphs* (they seem to fit somewhere on the stem of the dinosaur tree); examples include *Lagosuchus*, *Silesaurus* and *Herrerasaurus* (see Figure 19) but there are many more than just these three.

Careful analysis of the anatomy of many stem dinosauromorphs along with that of a variety of more clear-cut Triassic and Early Jurassic dinosaurs has revealed a new and unexpected genealogical pattern. The clade *Dinosauria* is maintained, but Seeley's traditionally hip-based split between Saurischia and Ornithischia has broken apart and it now appears that Theropoda and Ornithischia form a new clade *Ornithoscelida* (a name

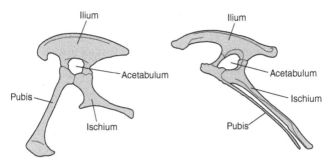

20. Traditionally the Dinosauria has been divided into two groups: Saurischia (left) and Ornithischia (right), because of the distinctively different arrangement of their hip bones.

originally invented by Thomas Huxley in 1870) that is the sister-clade to the Sauropodomorpha (see Figure 19) that represent all the generally very large, long-necked and long-tailed dinosaurs.

After such a long period of stability this new proposal will prompt lively debates and encourage further study. However, even at this early stage some new and potentially interesting evolutionary observations are apparent. For example *theropods* (and their descendants the *birds*) and *ornithischians*—the members of our new *ornithoscelidan* clade—both seem to have been capable of evolving a bird-like pelvic architecture, which is something that *saurischians* never achieved; it also appears to be the case that only *ornithoscelidans* evolved bird-like feathers.

Nearly dinosaurs

Herrerasaurus (see Figure 21) and its close relatives *Staurikosaurus*, *Chindesaurus* and *Sanjuansaurus*) along with *Lagosuchus*, and *Silesaurus* are representative of some very interesting and varied types of 'almost-dinosaurs' (dinosauromorphs) that ranged between herbivory, omnivory and carnivory in their diets and habits; they seem to have

21. Triassic 'near-dinosaurs' (*Herrerasaurus*) and examples of Triassic true dinosaurs: *Coelophysis* (an ornithoscelidan theropod) and *Plateosaurus* a (sauropodomorphan prosauropod).

explored a wide range of life-styles just prior to the emergence of true dinosaurs.

Sauropodomorphan dinosaurs

Sauropodomorphans are large-bodied animals that have pillar-like legs, very long tails, long necks ending in small heads, and their jaws tend to be lined with simple spatulate or peg-like teeth and were omnivores or herbivores. Some of the earliest

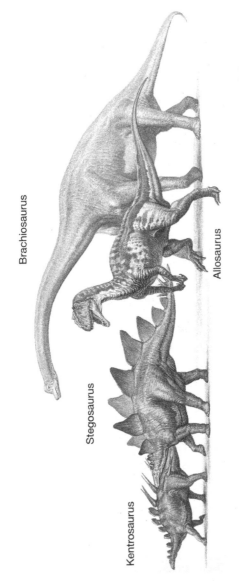

Brachiosaurus

Stegosaurus

Kentrosaurus

Allosaurus

22. Jurassic ornithischian thyreophorans: *Kentrosaurus* and *Stegosaurus*. The theropod *Allosaurus* and the sauropodomorphan *Brachiosaurus*.

(referred to as 'prosauropods'; see Figures 19 and 29, later in the volume), such as the Upper Triassic *Plateosaurus* ('flat reptile'—Figures 19, 21, and 29) were large (up to 10 metres long) and may have been omnivorous, rather than strictly herbivorous. Later Jurassic forms include veritable giants such as the diplodocoids (*Diplodocus* 'double-beam' and *Apatosaurus* 'headless reptile'—Figures 19 and 29), brachiosaurs (Figures 19, 22, and 29), and titanosaurians such as *Nemegtosaurus* that became abundant in the Cretaceous Period.

Ornithoscelidan dinosaurs

The new ornithoscelidan clade includes two anatomically diverse and abundantly represented sub-clades: *Theropoda* and *Ornithischia*.

Theropods superficially resemble herrerasaurids (and had previously been confused with them). Theropods are bipedal, predominately carnivorous dinosaurs (although some unusually specialized omnivores, insectivores, and herbivores evolved during the Cretaceous Period). A long, muscular tail counterbalances the front of the body at the hip, leaving the arms and hands free to be used to grapple with prey; their jaws are lined with sharp, knife-like teeth. Some early forms of this type include the small *Eoraptor* ('early predator'—Figure 19) and *Coelophysis* ('hollow face'—Figures 19, 21, and 29). Many theropods possessed skin that was covered by dense fur-like filaments or feathers.

Ceratosaurians ('horned reptiles') include early Jurassic examples such as *Dilophosaurus* and the later Jurassic *Ceratosaurus* (Figures 19, 22, and 29). *Allosaurians* include *Allosaurus* ('strange reptile'—Figures 19, 22, and 29) a large predator of the Jurassic and quite near relatives are the bizarre fish-eating *Baryonyx* ('heavy claw') and *Suchomimus*. Equally large and impressive are the tyrannosaurs (*Tyrannosaurus* and its close relatives) of the Cretaceous Period. Some of these theropods are already well-known to readers but the Theropoda, as a whole, is proving to be an extraordinarily varied group and in some instances quite bizarre-looking:

Ornithomimosaurians ('bird mimic reptiles'—Figures 19 and 29) resemble ostriches in their body proportions and the rather slender beaked jaws; however, some of their relatives, known as *therizinosaurs* ('scythe reptiles'—see also Figure 29) were huge, lumbering creatures with scythe-like claws on their hands, enormous abdomens, and ridiculously small heads; their jaws were lined with tiny teeth that are reminiscent of those of plant-eaters (hence their large bellies).

Oviraptorosaurians ('egg-stealing reptiles'—Figure 19) were lightly built, similarly ostrich-like animals whose heavy, snub-nosed heads had jaws that were toothless (and therefore had beaks as do living birds).

Dromaeosaurians ('running reptiles'—Figures 19 and 29) include *Velociraptor* ('high-speed killer') and *Deinonychus* ('terrible claw'—Figures 12, 19, and 29) and a host of similar but less memorably named creatures. Their skeletons resemble those of living birds and many of them are now known to have had feather-covered bodies; indeed, the similarities are so great that some dromaeosaurians are considered to be the ancestors of living birds (see Figures 19 and 29, and the next chapter).

Ornithischians are thought to have been herbivorous and, as with modern-day mammals, seem to be very numerous and far more anatomically varied than the predators that fed upon them. Several sub-groups are currently recognized.

Thyreophorans (Figures 19 and 29) are characterized by bearing armour-like bony plates in the skin on their heads and over their bodies; clubs or spikes may also adorn their tails. Their heavier build means that they are quadrupeds. *Stegosaurians* ('plated reptiles') named after the iconic *Stegosaurus* (Figures 22 and 29) well-known during the Jurassic for having tiny heads, a curious pelt of stud-like bones protecting their necks, rows of tall, angular bony plates along their back, and a spiky tail. *Ankylosaurians* ('welded reptiles') are also very heavily armoured Cretaceous forms including *Sauropelta* (Figure 19) as well as *Ankylosaurus* ('welded reptile') and *Euoplocephalus* ('truly plated head'). These latter were huge, and so heavily armour-plated that even their eyelids had bony shutters, they also had tails that ended in huge, bony clubs that could have been swung sideways to 'skittle' potential predators.

Cerapodans (Figure 19, see also Figure 29) were in contrast lightly built, unarmoured, and comprise the Ornithopoda, Pachycephalosauria, and Ceratopia. *Ornithopods* were medium-sized (2–15 metres long) and abundant (probably filling the ecological niches occupied by cattle, horses, antelopes, and goats today); some, such as *Hypsilophodon* ('high ridged tooth') were small (~2 metres long) balanced at the hip (just like theropods), had slender legs for fast running, grasping hands, and, most importantly, teeth, jaws, and cheeks adapted for a diet of plants. Throughout the reign of the dinosaurs, small to medium-sized ornithopods (2–3 metres long) were abundant. A number of larger types (ranging up to 15 metres) and known as iguanodontians (because they include animals such as *Iguanodon*—see Figures 19 and 29). Most impressive of all the iguanodontians were the hadrosaurian (duck-billed) dinosaurs of the Late Cretaceous (Figure 29). *Pachycephalosaurs* (Figure 29) had bodies that were very similar in general shape to ornithopods, but their heads were very odd-looking. The majority had a dome of thick bone on top that resembled a helmet. It has been suggested that these creatures were the 'head-bangers' of the Cretaceous world—perhaps using head clashing as part of their behavioural repertoire, as seen today among some cloven-hooved animals. *Ceratopians*

(Figure 29) include the fabled *Protoceratops* ('first horned face'—Figure 2) and the renowned *Triceratops* ('three-horned face'—see Figure 29). All had a characteristically narrow hooked beak at the tip of the jaws and tended to have a ruff-like collar of bone attached to the back edge of the skull, which in some instances grew to a huge size. While a few of the early ones maintained a bipedal habit, the majority grew in body size and greatly enlarged the head. Their great bulk and heavy head led them to adopt a four-footed stance, and their similarity to the modern-day rhinoceros has not gone unnoticed.

Box 2 The case of *Baryonyx*

The Early Cretaceous rocks of south-east England have been intensely investigated by fossil hunters (starting with Gideon Mantell) and geologists (notably William Smith) for well over 200 years. *Iguanodon* bones are very common, as are the remains of a limited range of other dinosaurs, such as '*Megalosaurus*', *Mantellisaurus*, *Barilium*, *Hypselospinus*, *Hylaeosaurus*, *Polacanthus*, *Pelorosaurus*, *Valdosaurus*, and *Hypsilophodon*. Given the intensity of such work, it would be thought highly unlikely that anything new would ever be discovered. However, in 1983 the amateur collector William Walker discovered a large claw bone in a clay pit in Surrey. This led to the excavation of an 8-metre-long predatory dinosaur that was entirely new to science. It was named *Baryonyx walkeri* in honour of its discoverer, and has pride of place on exhibition at the Natural History Museum in London.

The moral of this story is that nothing should be taken for granted; dinosaurs are only rarely preserved as fossils, so the fossil record is likely to be full of surprises.

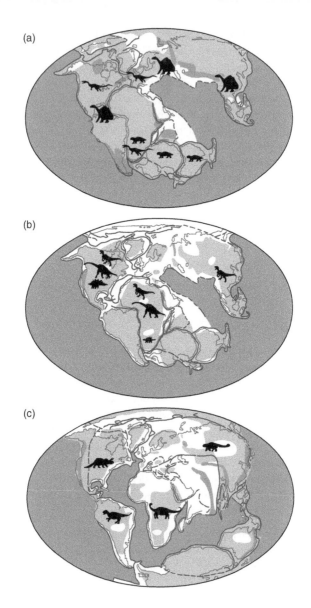

(d)

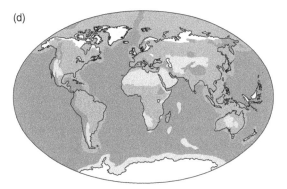

23. The changing continents. (a) Triassic Period showing the single supercontinent called Pangea. (b) Middle Jurassic Period. (c) Early Cretaceous Period. Note that the dinosaur images become increasingly different as the continents separate from one another. (d) The continents as they are today: close the Atlantic Ocean and the Americas fit neatly against West Africa.

As this all-too-brief survey demonstrates, dinosaurs were *many* and *varied*, judging by the discoveries made over the past 200 years. Even though to date ~1,000 genera of dinosaurs are known, this is a tiny fraction of the variety of dinosaurs that existed during the 170 million years of their reign during the Mesozoic Era. The majority will never be known: their fossilized remains may never have been preserved or were destroyed long ago. Nevertheless more will be discovered by intrepid dinosaur hunters in years to come (Box 2).

Dinosaur evolution and the Earth's changing geography

The dynamic (tectonic) Earth may well have influenced the overall pattern of the evolution of life. Evidence of the similarity of fossils collected from different areas of the world provided early clues that the continents might not have been as fixed in their positions as they appear to be today. For example, it was recognized that rocks and the fossils that they contained seemed to be remarkably similar on either side of the southern Atlantic Ocean. A small aquatic fossil reptile *Mesosaurus* was known to exist in Permian rocks in Brazil as well as in southern Africa. In fact as long ago as 1620, Francis Bacon had pointed out that the coastlines of the Americas and Europe and Africa seemed as if they could have fitted together (see Figure 23(d)), as if they were pieces of a gigantic jigsaw. On the basis of evidence from fossils, rocks, and general shape correspondence, Alfred Wegener, a German meteorologist suggested, as long ago as 1912, that in the geological past the continents of the Earth must have occupied different positions to the ones they are in today with, for example, the Americas and Eur-Africa nestled together in the Permian Period. Because he was not a trained geologist, Wegener's views were ignored or dismissed as unprovable speculations. Despite the appealing shape correspondence Wegener's theory simply lacked a mechanism: common sense told scientists that it was impossible to move things the size of continents around the solid surface of the Earth.

However, 'common sense' proved misleading. In the 1950s and 1960s very detailed maps of all the major continents, including their continental shelves, showed that they did indeed fit together remarkably neatly. Also many major geological features on separate continents became continuous when continents were fitted together. It seemed that such correspondence could not be accounted for simply by chance. Finally, magnetic surveys of the sea-floor found evidence that it was expanding in some areas and disappearing in others. The Earth's surface was actually divided up into huge tectonic plates that *moved* very slowly over time. The ocean floors, rather than being static, move like immense conveyor belts and the continents actually float upon these moving tectonic plates. The 'motor' that was driving this motion was the heat in the Earth's Core and the slow convection of heat-softened rock in the Mantle. The concept of *plate tectonics*, developed largely in Cambridge (UK), explains the otherwise puzzling observations of Bacon and Wegener.

From a dinosaur's perspective, the implications of plate tectonics are extremely interesting. Reconstructions of past configurations of the continents show that when dinosaurs first appeared on Earth (~230 Ma) all the continents were clustered together in a single giant landmass named *Pangea* ('all Earth'—Figure 23(a)). Dinosaurs could have walked all over the Earth, and the widespread discovery of similar-looking Triassic theropods and sauropodomorphs (Figure 21) confirms this.

During subsequent Periods: the Jurassic (Figure 23(b)) and Cretaceous (Figure 23(c)) Pangea gradually fragmented. The end-product of this process is the Cretaceous world (Figure 27(c)) that, though different geographically from today (note particularly the position of India in Figure 23(c)), has very recognizable continents.

A biological consequence of this process of continental sundering is that the once cosmopolitan Triassic dinosaurs became subdivided into isolated populations on different continents. Isolation is one of the keystones of evolution: once isolated, populations undergo evolutionary change in response to alteration in their local environment. In this instance, although we are dealing with comparatively huge (continent-sized) areas, each of the continental fragments carried its own population of dinosaurs (and associated fauna and flora). Each of these, with the passing time, evolved independently in response to changes stimulated by, for example, changing latitude, longitude, and prevailing climatic conditions.

Inevitably the tectonic events of the Mesozoic Era affected the overall evolutionary history of dinosaurs. Indeed, it seems perfectly reasonable to suppose that the fragmentation of ancestral populations over time must have accelerated the diversification of dinosaurs. Just as we can represent the *evolutionary* history of dinosaurs using family trees or cladograms, we can also represent the *geographic* history of the Earth through the Mesozoic Era as a series of branching events as continents separated from their 'parent' Pangea. Of course, this logic is a gross simplification of true Earth history because, on occasion, continents have coalesced, welding together previously isolated populations; equally, organisms may be able to hop (or swim or fly) between continents in some circumstances. But, as a first approximation, this concept provides a fertile area for investigating some of the larger scale events in Earth history and links biology directly to changes in the physical Earth.

Did the Earth influence ornithopod evolution?

The earliest work in this field of research was carried out by myself and a colleague (Andrew Milner at Birkbeck, University London), and concerned the ornithopods (Figures 19 and 29). Comparing the anatomy of a number of ornithopods a cladogram was constructed. To convert this into a phylogeny we plotted the known time ranges of species as well as their known geographic ranges. Some surprising patterns emerged that allowed us to build a general hypothesis about the evolutionary history of this group. We demonstrated that iguanodontians originated at a time of continental separation during the Late Jurassic: the ancestral population, from which both groups evolved, was divided by a new sea. Following isolation, one population steadily evolved into hadrosaurs in Asia alone, while the remaining iguanodontians persisted (but not as hadrosaurs) elsewhere; these two groups evolved independently through the Late Jurassic and Early Cretaceous Periods. However, during the latter half of the Cretaceous, Asia became reconnected to the rest of the northern hemisphere continents and its hadrosaurs were apparently able to spread across the northern hemisphere pretty much unhindered and seem to replace the non-hadrosaur iguanodontians wherever they came into contact.

While the pattern of replacement of iguanodontians by hadrosaurs in Late Cretaceous times seemed mostly consistent, there were one or two puzzling anomalies that needed to be investigated. There were reports, written at the turn of the 20th century, of iguanodontians in Europe (primarily from France and Romania) in rocks of very latest Late Cretaceous age. Our analysis did not expect iguanodontians to have survived into Late Cretaceous times because everywhere else the pattern was one of hadrosaur replacement. In the early 1990s, I re-studied the dinosaur fossils from Romania and France and showed that they did not belong to an iguanodontian, but rather represented an unusually long-lasting (*relict*) member of a more primitive group of ornithopods. So, one of the outcomes of our analysis was a great deal of new information about an old, but apparently not so well understood, dinosaur, and a suspicion that maybe Andrew and I had been correct in our hypothesis.

Similarly, a report published in the 1950s suggested that a very *Iguanodon*-like dinosaur lived in Mongolia in Early Cretaceous times. This tantalizing report also needed to be investigated to check whether its anomalous geographic range—in Asia in Early Cretaceous times—was real or, as in the Romanian example, another case of mistaken identity. The material, though fragmentary, was stored in the Russian Palaeontological Museum in Moscow, and I was again able to re-examine it. What emerged was not as we had predicted. This time the earlier reports were proved to be essentially correct, the genus *Iguanodon* (or something extremely similar to it) was present in Mongolia in Early Cretaceous times.

This second discovery did not fit so comfortably with the evolutionary and geographic hypothesis that Andrew and I had created. Indeed, in more recent years a suite of very interesting *Iguanodon*-like ornithopods have emerged in Asia, as well as North America, in what can best be described as 'middle' Cretaceous times at a time of the supposed mingling and spread of the hadrosaurs from Asia. These new discoveries suggest that our original evolutionary and geographic model although supported by some of the evidence may not have been correct in all respects: this is a classic example of an original hypothesis being subjected to testing by further research and new discoveries.

Dinosaurs: a global perspective

In more recent times, this approach has been applied much more broadly and in a more ambitious way. Two former students of mine, Paul Upchurch (University College London) and Craig Hunn, were encouraged to explore the entire family tree of the Dinosauria for evidence of similarities in patterns of geographic ranges, stratigraphic ranges and cladistic patterns of dinosaurs. It was an attempt to discover whether plate tectonics influenced the evolutionary history of *all* dinosaurs rather than just a few ornithopods.

Despite the inevitable 'noise' in the system resulting from the incompleteness of the fossil record and the confounding effects of the dinosaurs' ability to migrate between continental areas, it was pleasing to note that statistically significant coincident patterns emerged within the Middle Jurassic, the Late Jurassic, and the Early Cretaceous intervals. This suggested that tectonic events play *some* role in determining where and when particular groups of dinosaurs evolved. What is more, this effect has also been preserved in the stratigraphic and geographic distributions of other fossil organisms, so the evolutionary history of great swathes of organisms was affected by tectonic events and some of that imprint is still with us today. This is not new. I need only point to the unusual distribution of marsupial mammals (found only in the Americas and Australasia today), and the fact that distinct areas of the modern world have their own characteristic fauna and flora. What our research suggests is that we may well be able to trace the history of these distributions far more deeply than we had previously supposed was possible.

Chapter 5
Dinosaurs: their biology and way of life

Dinosaurs challenge our general understanding of how animals are constructed and how they work in real life because they seem to break the rules: they are larger than animals ought to be, certainly by comparison with what we see today, and they also seem odd, almost outlandish, in their design. Children of course find dinosaurs perfectly natural: 'weird' is normal as far as they are concerned, but researchers tend to be perplexed by the How? and the Why?

The following chapters illustrate attempts to understand the biology and way of life of dinosaurs.

Dinosaurs: hot-, cold-, or luke-warm-blooded?

As described in the Introduction, Richard Owen long ago speculated about the size, shape, and physiology of dinosaurs. Extracting meaning from the rather long-winded final sentence of his scientific report:

> The Dinosaurs...may be concluded to have...[a] superior adaptation to terrestrial life...approaching that which now characterizes the warm-blooded Vertebrata [i.e. living mammals and birds]. (Owen 1842, p.204)

The elephant-like reconstructions of dinosaurs that he created for the Crystal Palace Park (Figure 1) echoed his general view, but researchers at the time never grasped the physiological implications. Owen's daring leap of imagination was too advanced. Aristotelian logic prevailed: dinosaurs were anatomically reptilian, it therefore followed that they had scaly skins, laid shelled eggs, and, like all living reptiles, were 'cold-blooded' (*ectothermic*).

Thomas Huxley proposed, almost thirty years later, that birds and dinosaurs should be considered close relatives because of the shared anatomical similarities between living birds, the earliest known fossil bird *Archaeopteryx* (Figure 9) and the then newly discovered small theropod *Compsognathus* (Figure 10). He concluded that:

> it is by no means difficult to imagine a creature completely intermediate between *Dromaeus* [an emu] and *Compsognathus* [the dinosaur]…and the hypothesis that the…class *Aves* has its root in the Dinosaurian reptiles…. (Huxley 1868, p.365)

Huxley could also have asked: were dinosaurs then conventionally reptilian or were they more similar to 'warm-blooded' (*endothermic*) birds. However, it took another century *after* Owen's and Huxley's preliminary insights before palaeontologists began seriously to address these issues.

Owen's tentative connection between his 'dinosaurs [and the] warm-blooded Vertebrata' was interesting because the warm-blooded vertebrates of today include just the mammals and birds. Both groups are endotherms ('internal heat') because they generate their body heat biochemically and run all their bodily systems at a more or less constant temperature: they feel warm to the touch. This means that they can remain active on cold days. In contrast ectothermic ('external heat') reptiles: lizards, snakes, tortoises, and crocodiles rely on heat from the sun to control their body temperature: on a cold day, or when the sun does not shine, their bodies cool and they are obliged to become more or less inactive. The fact that endotherm bodies run at a constant temperature means that all the biochemical reactions that govern their activities work constantly and predictably. There is, of course, a cost to this style of life and that is that food has to be consumed and burned biochemically to create heat; and a large volume of air (oxygen) is also needed as part of that heat-generating (burning) process. Endotherms are costly to run (but work all the time) whereas ectotherms are cheaper to run (far less food and oxygen are required). Both regimes clearly work because there are animals that exhibit both styles of life today. The most obvious difference today is that endotherms (mammals and birds) exhibit far greater geographic distribution across our world (especially at high latitudes) and are far more numerous and diverse than the ectotherms, which are mostly restricted to lower latitudes.

So: were dinosaurs ectotherms? Endotherms? Or…did dinosaurs *exploit other types of metabolic regimes* that we have not considered?

As we saw earlier (Figure 13) Robert Bakker used the mixed fortunes of mammals and dinosaurs during the Mesozoic to prompt a re-think about dinosaurs as 'typical' reptiles. Their counter-intuitive fossil history begged the question: was there something 'special' about dinosaurs that allowed them to out-compete the mammals in the Early Jurassic? Was it because dinosaurs, as Owen and Huxley seemed to be hinting, were indeed endotherms and not typical reptiles at all?

Patterns of life on Earth

Living endotherms (mammals and birds) are not particularly good indicators of climatic zones because they are found everywhere, from equatorial to polar regions. Their endothermic physiology (and clever use of body insulation) allows them to operate more or less independently of prevailing climatic conditions. By contrast, ectotherms are reliant on ambient climatic conditions, and tend to be confined to warmer and sunnier (Mediterranean and tropical) zones.

Dinosaur fossils are in fact widely spread geographically; their remains have been found in Arctic and Antarctic land areas as well as in the tropics. Does this imply that they were capable of controlling their body temperature as modern-day endotherms? The easy answer would seem to be: yes. However, there is an important subsidiary question that should accompany the primary observation: what was the climate of Mesozoic Earth like? For example: were the polar regions, as we might expect them to be, cold and icy >66 million years ago? Rather interestingly the answer is…No! Mesozoic Earth was surprisingly warm, even at very high latitudes. There were neither Arctic nor Antarctic ice sheets, but instead these areas were covered with lush and diverse forests. Dinosaurs lived in a 'greenhouse' world that we would find not only strange but uncomfortable. This is an important lesson: we cannot simply assume that the world in the past was as it is today.

So, the discovery of high latitude dinosaur fossils does not necessarily *prove* that they were endotherms. Admittedly, during the winter seasons in the Mesozoic world, there would have been low light levels in the high latitudes, resulting in lower plant productivity: it seems that these high-latitude forests had a dramatic 'fall' as leaves were shed in the season of near-darkness. Perhaps the dinosaur fossils found at high latitude were actually the carcasses of unfortunate summer migrants to the high latitude pastures. In contrast to the winter period the sun would have shone for nearly twenty-four hours a day and plant growth would have been at a maximum during the summer period. Frustratingly this latter scenario would be far harder to prove (or falsify) because we do not have sufficient time resolution in the terrestrial geological record to distinguish between summer and winter rocks.

The machinery: legs, heads, hearts, and lungs

Owen noted that dinosaurs place their feet vertically beneath the body on straight 'pillar-like' legs for weight support; and, as we have seen, such legs could also be flexed 'springy' legs for fast running in some varieties. The only living animals that adopt these types of posture are birds and mammals (both of which are endotherms). All the other vertebrate groups 'sprawl' with their legs directed sideways from the body (and are, of course, ectotherms).

We have seen that some dinosaurs were light, slender-limbed, and apparently built to move quickly. This observation appeals to our common sense: Nature tends not to do things unnecessarily. If an animal is built as if it could run fast, it probably did so. It is perfectly reasonable to expect such a creature had a high-powered 'engine' to drive fast movement. Does that also point to their being endothermic? Yet again we need to be slightly cautious with this line of reasoning because it is known that living ectotherms can move *extremely quickly* indeed—crocodiles, alligators, and Komodo dragons can pounce on and kill unwary prey (including humans). The crucial factor is that ectothermic crocodilians and Komodo dragons cannot *sustain* fast movement: their muscles build up a large oxygen debt very quickly and these animals have to rest after short bouts of very energetic activity so their muscles can recover. Endotherms, by contrast, can move quickly but have *endurance* because their high-pressure blood system and efficient lungs very quickly replenish the food and oxygen needed by their muscles.

A subtle refinement of the discussion about upright limbs is the observation that the ability to walk *bipedally* is linked exclusively to endothermy: some mammals and all birds are, and many dinosaurs were, bipedal. This reasoning requires some thought about how a bipedal posture is maintained. A quadruped has the advantage of considerable stability when it walks. A biped is inherently unstable and to walk successfully sensors monitoring balance, fast-twitch muscles to correct and maintain balance, and a rapid coordinating system (the brain and central nervous system) are essential if this posture is to be maintained. The functionality of the brain is central to this whole instability problem and it must be able to work quickly and efficiently at all times. This implies that the body is able to provide the brain with constant supplies of oxygen, food, and heat to allow its chemistry to work optimally. The prerequisite for this type of coordination is a constant endothermic physiology. Ectotherms periodically shut down their activities when cold, which makes the brain less responsive to body functions and would be incompatible with a bipedal life style.

Another posture-related observation can be linked to the efficiency of the heart and its potential to sustain high activity levels. Many birds, mammals, and dinosaurs adopt(ed) a body posture in which the head is normally held at a level that is appreciably higher than that of the heart. This difference in head-heart level has a hydrostatic consequence: the heart has to be able to pump blood at *high* pressure 'up' to the brain. However, blood is also pumped from the heart to the lungs with each heartbeat and must circulate through the lungs at *low* pressure, otherwise it would burst the delicate capillaries that line the respiratory membranes in the lungs. To permit this pressure difference, the heart in mammals and birds is divided down the middle, so that one side of the heart (the *systemic*, head and body, circuit) is larger and more muscular so it can send blood at higher pressure than the other side (the *pulmonary*, or lung, circuit).

Living reptiles carry their heads at roughly the same level as the heart nearly all the time. Their hearts are not divided into high- and low-pressure halves like those of mammals and birds so they pump blood around their bodies at lower pressure. Interestingly, the reptilian heart and circulation offers advantages for these creatures; they can shunt blood around the body in ways that mammals and birds cannot. For example, ectotherms spend a lot of time basking in the sun to warm their bodies.

While basking, they can preferentially shunt blood to the skin, where it can be used to absorb heat (rather like the water in solar panel heating pipes). The major disadvantage of this system is that their blood cannot be circulated at high pressure—a feature that is essential in any animal that is moving quickly and needs to bring food and oxygen quickly to its hard-working muscles.

Linked to the effectiveness of the heart and circulatory system there needs to be an ability to supply sufficient oxygen to muscles to allow high levels of aerobic activity. In some groups of dinosaurs, notably the theropods and the giant sauropodomorphs, there are anatomical features concerning their lung structure. In both these groups of dinosaur there are traces of pouches or cavities (called *pleurocoels*) in the sides of the vertebrae along the backbone. In isolation, these might not have attracted particular attention; however, living birds show similar features that equate with the presence of extensive thin-walled diverticulae of their air-sacs: because of this pervasive air system birds are often described as *pneumatic*. Air-sacs and their diverticulae reduce the weight of the bird (and dinosaur) skeleton and are, in birds, associated with a bellows-like mechanism that permits birds to breathe very efficiently. It seems that some dinosaurs (particularly the theropods) may have had a bird-like respiratory system.

The implication of these observations and inferences is that some dinosaurs, because of their posture and locomotor style, needed a fast and responsive neuromuscular system and large brain; this system would have in turn required a high-pressure blood system and efficient lungs. All these features are compatible with the high and sustained activity levels that are found today only in living endotherms.

Predatory life-style

Although the line of argument that follows is not applicable to all dinosaurs, it is instructive in the sense that it shows what *some* dinosaurs were capable of doing. As John Ostrom demonstrated (Chapter 2), *Deinonychus* (Figure 12) was a sharp-eyed predator that could run fast; had an unusual stiff, narrow tail; extraordinary gaff-like inner toes on its hind feet; and long, sharply clawed, grasping arms. Its skeleton was also pneumatic, so its body was light and probably had efficient lungs. It is not unreasonable to suggest that this animal was built as a pursuit predator, capable of using its narrow tail as a balancing aid and it very probably leapt on its prey, which it then mortally wounded using the 'terrible claws' on its feet.

We have never seen *Deinonychus* in action, its predatory style of life is an interpretation based on observable features of the skeleton. However, one truly remarkable fossil discovered in Mongolia supports these conjectures. The latter comprises two dinosaurs: a small herbivorous ceratopian *Protoceratops* (Figure 2) and a *Velociraptor* (a very close relative of *Deinonychus*). This extraordinary fossil (displayed in the dinosaur museum of Ulaan Baatar) shows these two creatures caught in a death struggle; they very probably choked to death in a dust storm while fighting each other. The *Velociraptor* is preserved clinging to the head of its prey using its long arms and is in the act of kicking at the throat of its unfortunate victim.

Such 'sophistication' in the design of such killing machines as *Deinonychus* and *Velociraptor* suggests activity levels more similar to those displayed by modern endotherms: modern raptors (predatory birds) and large felines (members of the cat family), rather than typical reptiles.

Bone histology

In recent years considerable attention has been focused on understanding the internal structure of dinosaur bone. The mineral structure of fossil bones is largely unaffected by fossilization and it is possible to cut very thin sections of dinosaur bone that can be examined microscopically. These show that dinosaur bone was highly *vascularized* (penetrated by many tiny blood vessels) and closely resembled sections taken from living endothermic mammals. Highly vascularized bones can be formed for different reasons. For example, one pattern of vascularization (*plexiform* or woven bone) reflects the rapid deposition of bone crystals (*apatite*) on a web of collagen and dense capillaries on the surface of a growing bone; this is particularly associated with young animals that are growing quickly. This somewhat loose aggregation of minerals is progressively replaced by secondary (*Haversian*) bone that represents a phase of strengthening of the woven bone by internal remodelling that occurs a little later in the life of an individual.

What can be said is that many dinosaur remains show evidence of them being able to deposit bone minerals fast, and to grow quickly, linked to the ability to strengthen these bones by internal remodelling. Dinosaurs sometimes exhibit periodic interruptions in their pattern of growth (which mimics the intermittent—seasonal—pattern seen in the bone growth of living reptiles), but this style of growth is by no means uniform. Equally, and less probably, some endotherms (both bird and mammal) exhibit a style of bone structure (*zonal*) that displays very little vascularization, while living ectotherms can exhibit highly vascularized bone in parts of their skeletons.

There are, surprisingly, no absolute correlates between an animal's physiology and its internal bone structure. It seems that taking just one or two thin sections can be very misleading. What is needed is a very detailed set of thin sections at different orientations within single bones, and also across many bones of the same skeleton in order to build up an accurate picture of the way any dinosaur grew its bones, and how similar or different this might be to living endotherms (or ectotherms). It is certainly the case that *some* dinosaurs grew their bones in the same way as living endotherms—but *not all* did!

Size matters: go large!

Dinosaurs were, on average, large animals: the smallest (excluding hatchlings, of course) were about 1 metre or so in length, while the largest were 30+ metres long. The average across all dinosaurs was in the 3–5 metre range, which is an order of magnitude larger than the 30 cm average for all living mammals. But why does size matter?

Small animals have a comparatively large skin surface when compared to large animals. The reason for this is that for every doubling in length of the body, its surface area increases by the square of that linear measurement, whereas its volume (therefore the amount of tissue, and therefore its mass) increases by the cube of the linear measurement. So, as an animal increases in size its surface area reduces compared to its volume/mass. This phenomenon has important consequences, especially in relation to the ability of any animal to retain heat or, in effect, control its internal body temperature. The larger the animal, the smaller its surface area (across which it can lose body heat), and the more stable its internal body temperature. This effect has been neatly shown by simple experiments carried out using young and adult alligators. Measuring the body temperature of the young, using implanted thermocouples, showed that across a day–night cycle the body temperature exactly followed the ambient air temperature. In contrast, adult alligators conserved body heat and their day–night temperature changes were far more modest: the larger adults were effectively buffered against external temperature change because they did not lose so much heat through the skin across the day–night cycle.

Just using that simple observation (young alligators are a mere 15–20 cms long, while adults are 3–4 metres in length) alligators, even adults, large though they seem to us today are comparatively small beside the majority of dinosaurs, and are also rather elongated and cylindrical in body proportions (which would tend to increase their general body surface area).

So how does this translate into our understanding of dinosaurs in the Mesozoic world? It has already been established that the Earth was considerably less latitudinally zoned and much warmer than is the Earth today. So dinosaurs lived in a continuously warm world. Being very large animals, dinosaurs would have tended to be comparatively thermally buffered: their core body temperatures would not have fluctuated to any great extent because their body surface area was relatively small in comparison to the volume of their tissues. This means that dinosaurs were not only warm, because the ambient temperature was warm, but they also had the virtue of being thermally stable because of their large size. Some physiological functions, perhaps most particularly the heat produced by the large muscles that both supported their skeletons and moved them around, would have added supplementary heat that would have allowed their biochemistry to operate close to optimally much of the time. In the case of the *very* large herbivores (sauropodomorphs) the huge fermenting chambers for the digestion of plant tissues within their guts would also have generated useful chemical heat. Dinosaurs were in many ways rather fortunate: they lived in just the right sort of greenhouse world!

Elephants: do they help us to understand dinosaurs?

If the line of reasoning above has merit then it may be the case that very large dinosaurs (notably the multi-tonne sauropodomorphs) may have been 'embarrassed' by their heat production: they may have run the risk of over-heating—which can be fatal. This might explain some of their otherwise unique anatomical features. Modern elephants live in hot tropical environments and are known to suffer from heat-stress on occasion. Elephants generate internal body heat, they are genuine endotherms, and can be quite active animals. In hot climates, muscular heat (generated by postural and locomotor muscles) increases an elephant's overall heat production. They also have a large body volume and a comparatively small skin surface area across which to lose heat. Being hairless, of course, helps them to radiate body heat through the skin, but that is insufficient for their overall heat-budget. Over-heating is a very real (and potentially fatal) problem. The solution to that problem is to have very large ears! Elephants can pump large volumes of hot blood through their enormous ears, and these can be actively flapped, if necessary, to create a breeze that helps them to radiate heat very effectively. Is this observation relevant with respect to large dinosaurs?

The core part of the body of any large sauropodomorph (excluding the very long tail and neck) is very elephant-shaped (and not too much bigger than some of the largest known extinct elephants). Such enormous animals would, like elephants, have produced considerable quantities of muscle-generated heat (irrespective of whether they were endotherms or ectotherms). They also lived in a world that was substantially warmer than today—estimates suggest that mean global temperatures may have been as much as 8–10°C higher. Sauropod heat-balances may have been a real physiological problem. There are two features that might have assisted with this problem (there is no evidence for them having huge, flappable ears!): firstly, their long cylindrical tails and necks may well have acted as large-surface-area radiator structures that helped them to off-load some body heat; secondly, they had large, complex air-filled passages (pneumatic cavities and diverticulae) within their backbones, their ribs, and upper limb bones that may have acted as a heat-venting system. One can almost imagine that the exhaled breath leaving the mouth of a large sauropod had some of the characteristics of the exhaust from an internal combustion engine: that is to say HOT, with plenty of CO_2 and some water vapour, but without the more damaging carbon monoxide, nitrogen, and sulphurous hydrocarbon supplements.

Are we close to understanding dinosaur physiology?

Using the example of super-large dinosaurs makes a general point about how scaling effects can influence the internal environment of the body of any dinosaur. Not all dinosaurs were so big, the majority were animals whose body size lay within the 4–12 metre body-length range. As is known to be the case in large crocodilians today, these animals would have been substantially thermally inert, but also had the benefit of living at a time when the world was far warmer than it is today. Therefore their body temperature may quite reasonably have been controlled within fairly narrow limits (whether they were physiologically endothermic or ectothermic). They could have been ectotherms that were constantly warm-blooded (like an endotherm) without any of the metabolic costs of being genuinely endothermic: a condition that could be described as *ectothermic homeothermy* (or as some have called it 'gigantothermy')—*a perhaps uniquely dinosaurian solution to living in the hot Mesozoic world.*

Might dinosaurs have had differing physiologies?

Although the majority of dinosaurs were large (considerably more than 1 metre long) there are two notable types of dinosaur that bucked this overall trend: baby dinosaurs and the dinosaur ancestors of true birds.

Baby dinosaurs hatched from eggs, we know this because they are preserved as fossils on occasion, some eggs even have embryonic dinosaur remains inside that give us a glimpse of early stages in their growth. Dinosaur eggs range in size from something that is roughly equivalent in size to a grapefruit up to that of a basketball.

Hatchlings can only have a body size equivalent to that which can be folded into their respective eggs; this implies that the majority would likely have been much less than 1 metre in body length. Comparatively small dinosaurs would, therefore have had a large skin surface, relative to their body volume and would have tended to lose body heat to their environment quite quickly (especially when in the shade, or on a cool, cloudy day—conditions that might occur even in warm Mesozoic world). Their physiological options would have been limited: either their body temperatures varied with changing air temperature, or they were in some way stabilized to avoid over-cooling and potential death. Endotherms today incubate their young, keeping them snug and warm in nests, and using parental body heat to assist their heat balance—this might have been an option for some of the smaller adult dinosaurs, but it is hard to imagine a 15–30 tonne sauropod snuggling up with its young: their nests would have needed to be unimaginably big, and the risk of crushing their young far too great.

Alternatively it seems at least likely that the breeding season was such that the young emerged during the warmest part of the year and suffered minimal cooling when they were at their smallest and most vulnerable to heat-loss. Perhaps linked to this, the structure of their bones when studied histologically, suggests that youngsters grew exceptional quickly (as do modern birds and mammals) so they reached a body size, which made them less susceptible to cooling not long after hatching. Finally of course there is the possibility that such youngster kept themselves warm by being insulated (having a covering of hair-like filaments or down-like feathers). This thought brings us to the consideration of naturally small-sized dinosaurs and some remarkable recent discoveries.

Naturally small *ornithopods* and *ceratopians* (see Figures 19 and 29, earlier and later in the volume, respectively) are known (~1 metre long) but there is only one group of dinosaurs that display a consistent overall trend toward small size (1 metre down to tens of centimetres in body length) and these are the *avian-theropods*—(such as *dromaeosaurians*—Figures 19 and 29)—so-called because they seem most closely similar to living birds, or *avians*. This small-size trend is curious: dinosaurs are almost by definition *large* animals. Most living birds (there are exceptions of course, think of ostriches) are small and light in order to be able to defy gravity and fly. Living birds are also known to be endotherms: they generate internal body heat to keep their bodies warm and at a constant internal temperature (despite their small size and the correspondingly large surface area of skin) by being insulated using a covering of feathers. So, the obvious question: how much can we know or deduce about avian-theropods: the dinosaurian bird ancestors?

Many small theropods (*Compsognathus*—Figure 10) do have skeletons that are quite bird-like—they are light and delicate, have long slender legs, a long neck, and large, forward-pointing eyes. However they also retain many more obviously 'reptilian' or dinosaurian features such as the clawed hands, teeth in the jaws, and a long, bony tail.

Deinonychus: the avian-theropod template

The skeleton of *Deinonychus* (Figure 12) displays a tail that is extremely narrow and stiffened by bundles of long, thin bones, the only flexible part being close to the hips. A thin non-muscular tail dramatically changes the pose of this animal because it is no longer heavy enough to counter-balance the front half of the body. Without other anatomical changes this dinosaur would have constantly pitched forward on to its nose.

To compensate for the loss of the heavy counter-balancing tail, the skeletons of these theropods were re-engineered: the *pubic* bone, which marks the rearmost part of the gut, normally points forward and downward from the hip socket. However this bone is rotated backwards to lie parallel to the ischium (the other lower hip bone)—in a truly bird-like rearrangement. This permitted the belly to shift backward beneath the hips; this shift compensated for the loss of the counter-balancing tail. Another subtle change was to shorten the chest in front of the hips. The chest is also stiffened to anchor the long praying-mantis-like arms and gaff-clawed hands.

Archaeopteryx, bird or dinosaur?

Archaeopteryx (Figure 9) exhibits many avian-theropod features: its tail is long and thin; its hip bones are arranged so that the pubis points backward and downward; the jaws are lined with small, spiky teeth (not a typically bird-like beak); its arms are long and jointed so that they can be extended and folded (like a praying mantis) just as in avian-theropods, and the hands have three sharply clawed fingers that are identical to those seen in *Deinonychus*.

Specimens of the Late Jurassic (~150 Ma) *Archaeopteryx* are remarkable because they were buried in lake sediments that permitted feather impressions to be preserved on its wings and along the sides of its tail. As a result this fossil creature was easily classified as the earliest known bird. Feathers had been, since the 18th century, regarded as *the* unique defining characteristic of birds.

Had chance not led to the preservation of feathers it is tempting to wonder how this creature might have been classified. It might well have been classified as an unusually long-armed, small theropod dinosaur.

Chinese wonders: deeper mysteries from the Orient

Quarries in Liaoning Province in north-eastern China began, during the 1990s, to yield some quite extraordinarily well-preserved, fossils of Early Cretaceous age. At first, these comprised beautifully preserved early birds such as *Confuciusornis* (Figure 19) whose skeletons often included remnants of feathers, beaks, and claws. Then in 1996, a complete skeleton of a small theropod dinosaur, very similar in anatomy and proportions to the theropod *Compsognathus* was described by Ji Qiang and Ji Shu'an and named *Sinosauropteryx* (Figure 24) and displayed soft tissue remains in the eye socket, the belly area, and, around the body margin, a halo of coarse filaments.

Shortly after *Sinosauropteryx* was discovered another skeleton was revealed. This animal, named *Protoarchaeopteryx*, was the first to show the presence of true bird-like feathers attached to its tail and along the sides of its body, and its anatomy was much more similar to that of a smaller version of the avian-theropod *Deinonychus* than *Sinosauropteryx*. Then another discovery revealed an animal that was almost identical to *Deinonychus*, but this time named *Sinornithosaurus* (and around its skeleton were patches of densely packed short filaments). Each new discovery intensified the search for more specimens. In the spring of 2003 a remarkable 'four-winged' avian-theropod (*Microraptor*) was described. This creature was small yet similar to *Deinonychus*, However, what was striking was the preservation, along the arms, of feathers creating *Archaeopteryx*-like wings and, very unexpectedly, wing-like fringes of elongate feathers attached to the lower parts of each leg—hence its entirely understandable nickname *four-wing*. As time has gone on and more discoveries have been made, including several more four-winged theropods, so it seems that feather-legged avian-theropods were not at all unusual at this time in Earth history.

The avalanche of new, and seemingly ever more startling, discoveries from the quarries in Liaoning over such a short space of time make it is almost impossible to imagine what might be discovered next. Looked at from a distance it would seem that the development of an insulated covering to the skin (initially

24. *Sinosauropteryx*. The almost complete skeleton of this small theropod dinosaur collected from a quarry in Liaoning, north-east China, preserves around its skeleton a halo of dark-staining tissue that indicates there was some sort of skin covering that looked rather like dense hair-like fibres. There are also remnants of organic material in the area of the eye socket and in the belly region. Such preservation is very exceptional.

filaments, later on downy feathers, and later still longer contour and flight feathers) opened up an entirely new range of evolutionary opportunities for small predatory dinosaurs across the interval between the Jurassic and Early Cretaceous. It seems that we are randomly sampling part of a remarkable evolutionary 'bomb-burst' of small, very active feathery creatures.

Lately Xu Xing and colleagues in Beijing and Nanjing described a yet stranger theropod, *Yi qi* (Ee-chee: 'strange wing'). The preserved remains, described in 2015, and collected in Hebei Province in China (just south and east of Liaoning Province) are of Middle Jurassic age and pre-date *Archaeopteryx*. However the skeleton, although it resembles other small, avian-theropods has some very peculiar features. The skull is robust and short-snouted and teeth line the jaws, the neck and chest are comparatively short as is the tail, which was festooned with a pair of 'streamers' made of elongate feather-like filaments. The legs are very long and slender, but the arms are even longer and the fingers of the hand are greatly elongate, and there is an additional bony rod extending outward from the wrist (called a *styliform element* for want of a better name). The body is covered by oddly shaped feathers that have narrow stems and branching clusters of finer filaments near their tips (giving the animal a very furry appearance in life). However, the most unexpected discovery was a thin membranous sheet of skin on the arm: the animal appears to have had a bat-like wing membrane, supported by the elongate styliform element in addition to its feathery body covering. Trying to explain and describe how this animal lived is quite a challenge because the combination of feather-like structures and thin membranes is so unexpected.

Now that membranous wing structures have been recognized (and accepted as real, rather than as weird preservational artefacts) it has been realized that other feathered avian-theropods such as *Caudipteryx, Anchiornis*, and *Microraptor* also bear traces of this membranous wing (or bat-like 'patagium'). Such discoveries emphasize how little we know, and how many assumptions have been made about bird origins and evolution.

Experimental bird-like dinosaurs

It is clear that in Middle Jurassic times, while the majority of dinosaurs were very large and imposing creatures, there was an extraordinary diversity of small, highly agile, and active bird-like theropod dinosaurs. These latter were clearly 'experimenting' with body design and functionality to an extraordinary degree.

Climbing may well have been one of these dinosaurs' activities. Developing thin sheets of skin stretched between the front and hind limbs that could act as simple parachutes or gliding surfaces to break their falls, or to assist leaps from tree to tree (similar structures are seen today among a range of tree-dwelling mammals such as the so-called 'flying squirrels' and 'sugar gliders'). Feathery skin structures would have provided insulation, but evolution may well have acted to 'select' those individuals that possessed rather more flamboyant feathers, so that they became not only larger, but perhaps also coloured to improve the sexual attractiveness of individuals. Considerably later, elongate feathers started to take on the ability to act as aerofoils for acrobatic displays, and ultimately served as flight feathers that are associated with wings (thereby replacing the somewhat more delicate and probably vulnerable skin membranes). It seems that feathers designed specifically for flight evolved quite late on in the history of avian-theropods.

Insulation for all?

The preservation of delicate structures on the surface of the skin of theropod dinosaurs has transformed what were previously 'speculations' about the physiology of small dinosaurs into far more strongly supported interpretations. It is intriguing to be able to report that skin surface structures, which look like quills and simple-tufted feathers, have also been reported in small ornithischian dinosaurs such as *Psittacosaurus*, *Tianyulong*, and *Kulindadromeus* (Figure 25).

The importance of these latter discoveries is that it implies that ALL dinosaurs had the *potential* to grow quills, filaments, or feather-like structures in their skin. This does not necessarily mean that all of them did. There is in fact a wide range of examples of dinosaurs (theropod, sauropodomorph, and ornithischian) whose skin impressions have been preserved and these show tessellated patterns of small scales—very similar to the skin texture seen in modern reptiles (and as can also be seen on the legs and feet of living birds). Clearly being able to grow such structures can be useful in some circumstances, but not in *all* circumstances.

To date no *sauropodomorph* remains have revealed bristles, filaments, or protofeathers. Skin impressions are known for a few

25. *Kulindadromeus*. This comparatively recently described ornithischian from Russia exhibits a variety of skin textures, ranging from typical reptilian scales on its tail and legs, filaments forming a hair-like coat, and in places short and frilled more protofeather-like structures on its upper arms and legs.

sauropodomorph taxa and these show simple skin-scale patterns. Indeed it is quite difficult to quite imagine a huge multi-tonne sauropod wearing the equivalent of a fluffy overcoat! Nevertheless, such structures, though perhaps unlikely for physiological reasons, cannot entirely be ruled out across any of the recognized dinosaurian groups. The reason for making this sweeping statement is rooted in logic.

We consider that the nearest relatives of dinosaurs (and by that I include living birds), within the group known as archosaurs, are the *pterosaurs*, (the 'pterodactyls' or flying reptiles of the Mesozoic Era). It has long been known that pterosaurs had a short dense, fur-like skin covering; this makes perfect sense because pterosaurs were flyers, and their small bodies and membranous wings would have cooled very quickly in the air, especially at altitude. Pterosaurs would have *needed* a constantly warm body to maintain brain function and normal bodily activities whilst flying (otherwise they would have lost control of their muscles and wings, and simply crashed to Earth). If we allow that pterosaurs were the sister-group to dinosaurs (and birds) then we can state with confidence that not only pterosaurs could grow filaments/fur, but so could all their 'cousins'. So the entire range of dinosaurs, and of course their descendants the birds, would have shared the biochemical machinery necessary for their growth. Whether particular groups did so would depend on their particular styles of life and biological needs. It is a little bit like saying that *all* mammals are covered in hair...well, yes, many are, but there are mammals that get along perfectly well without hair or fur. Humans have hardly any hair compared to their primate relatives; elephants are nearly naked; naked mole-rats, well...it's in the name.

Looked at logically it is a little curious that palaeontologists regarded the discovery of filaments and feathers on some dinosaurs as a surprise: their presence was entirely predictable.

The debate: dinosaur physiology, bird flight, and behaviour

It is now obvious that our predecessors were not correct: feathers do not, after all, make a bird. Various sorts of skin covering were present across a wide range of dinosaurs, from a shaggy, fur-like covering, to downy and even properly *vaned* (contour and flight-capable) feather structures. These discoveries force us to wonder just how widespread such body coverings might have been. Is it not reasonable to wonder whether giants such as *Tyrannosaurus rex* had some sort of epidermal covering...even if only as nestlings? Such tantalizing questions cannot be answered, they require new fossils from geological deposits that allow soft-tissue preservation.

From a physiological perspective, however, the proof of the existence of dinosaurs with some sort of insulation indicates that these dinosaurs were endothermic. There are two reasons for believing this:

1. Many of these feathered dinosaurs were small-bodied (20–40 cm long) and, as we know, small animals have a relatively large surface area and lose body heat to the environment very quickly. Therefore insulation using filaments (which mimic the fur seen on the bodies of living mammals) or downy feathers was a necessity because these creatures were generating internal body heat.

2. Equally, the possession of an outer layer of insulation on the skin would have made basking difficult, if not impossible, because this layer would have inhibited their ability to absorb heat from the sun. Basking is the ectotherm's way of gaining body heat, so a furry/feathered lizard is physiologically improbable.

Body coloration and its link to an animal's physiology

Rather intriguingly there is an additional reason for believing that insulated and very bird-like dinosaurs were endothermic—this comes from detailed study of the preserved filament and feather structures themselves. It has been noted that areas of some of these structures appear to have lighter or darker shading, suggestive of ancient colour patterns. The shading is caused by the presence of tiny dark-staining structures that correspond in size and shape to *melanosomes*, which are responsible for the colours in the skin, hair, and feathers of living animals. Detailed study of these fossilized melanosomes has permitted the tentative reconstruction of some colour patterning for these animals. While most of the estimates of colour suggest greens, browns, and blacks, notably for the filament-covered dinosaurs, it has been noted that melanosomes become more varied in structure in fossils that have vaned feathers. A greater variety of melanosomes is seen in living endothermic mammals and birds, and seems to reflect their endothermic status as well as their greater colour variability. So, the implied colourfulness of feathered dinosaurs and early birds may link directly to their having been endothermic.

In summary: small dinosaurs...

The implications of these new discoveries are truly fascinating. Small dinosaurs developed insulation to prevent heat loss because they were likely endotherms—not because they 'wanted' to become birds! The early Chinese discoveries indicate that various types of insulatory covering developed ranging from hair-like filaments to full-blown feathers. Bird flight did not evolve immediately but had a far more prosaic origin. Several avian-theropods from Liaoning seem to have tufts of feathers on the end of the tail (rather like a geisha's fan) as well as fringes of short feathers along the arms, legs, on the head, or running down the spine. Preservational bias may dictate how and on which parts of the body these may be preserved; however, it seems that more recognizably shaped feathers (with a broad vane) evolved as structures linked to the behaviour of these animals—providing colourful recognition signals, as with living birds or as part of their mating rituals, long before any genuine flight function had developed.

Gliding and flight, rather than being the *sine qua non* of avian origins, become later, 'add-on' benefits. Obviously, feathers have the *potential* for aerodynamic uses; just as with modern birds, the ability to jump and flutter may well have embellished 'dinobird' mating displays. For example, in the case of the small avian-theropods a combination of fringes of feathers along the arms, legs, and tail might have provided them with the ability to launch into the air for a brief aerobatic display from branches or equivalent vantage points. From just this sort of starting point, gliding and true flapping flight seem a comparatively short 'leap' indeed.

Bird-like behaviour preserved in fossils

Just to reinforce the idea that some dinosaurs were not only bird-like in shape, a few fossils exhibit bird-like behaviours as well. In 2004, a report appeared in the journal *Nature* that suggested small bird-like avian-theropod (a *dromaeosaurian*—Figure 19—named *Mei*) displayed a very bird-like sleeping or resting posture with its body resting on folded legs, its front limbs folded neatly against its sides and its neck and head curved backwards and tucked under one arm (wing?). It seems that only mammals and birds today rest/sleep with their limbs folded, and only birds tuck their heads in that manner. The 'tucked-up' pose minimizes its surface area and therefore conserves body heat effectively (especially its head and warm brain, which would otherwise lose a great deal of heat). The fossil preserves no feathery skin covering, but it seems likely that such a small, delicate creature would have been both endothermic and insulated—as in modern birds.

Another remarkable fossil from Mongolia reveals even more explicitly bird-like behaviour. *Citipati* is the name given to an *oviraptorosaurian* (Figure 19) avian-theropod whose partial remains were found with a group of eggs. Paradoxically 'oviraptor' means 'egg-thief' because it was at first thought that these dinosaurs stole eggs from the nests of other dinosaurs—well, perhaps they did (who can honestly tell?), but this fossil shows quite the reverse. The animal is preserved crouching with its arms encircling a nest of eggs as if it were not only protecting, but incubating them—just like a modern bird.

Persistent concerns

Earlier in this chapter it seemed perfectly reasonable to conclude that dinosaurs lived at a time in Earth history that favoured large-bodied, very active creatures that were able to maintain a stable, high body temperature without most of the costs of being genuinely endothermic. However, the existence of small, insulated dinosaurs suggests that this view cannot be entirely correct—small, insulated dinosaurs *simply had to be endothermic*. The close relationship between theropods and living birds (the latter being indisputably endothermic) reinforces that view.

Most dinosaurs were very large and their bodies would have been capable of maintaining a constant internal temperature. Extrapolating from the elephant, it would not have been in large dinosaurs' interests to be genuine endotherms in a world that was in any case very warm. Having evolved physiologically as ectothermic homeotherms (having a stable internal body temperature that was made possible by large body size *and* because they lived in a greenhouse world), the only groups of dinosaurs that bucked the general dinosaurian trend toward huge size were small-bodied ornithischians and dromaeosaurian theropods.

It is clear, from their anatomy alone, that small active dinosaurs would have benefited from being homeothermic (running their body at a more or less constant temperature). Paradoxically, homeothermy cannot be maintained at small body size without an insulatory covering because of the otherwise uncontrollable heat loss through the skin. In a sense the choice was stark and simple: small dinosaurs had to either abandon their high-activity lifestyle and become conventionally reptilian (shutting down activities when cold), or boost internal heat production and become properly endothermic, and control excessive heat loss by insulating the skin surface.

Best of both worlds?

So, I propose that dinosaur physiology was not a case of 'all or nothing': most dinosaurs were greenhouse-world dwelling ectothermic homeotherms, perfectly capable of sustaining high activity levels without having to bear the full cost of mammalian or avian styles of endothermy. However, very small dinosaurs (and their descendants, the true birds) were obliged to develop full-blown bird/mammal-like endothermy.

Chapter 6
Dinosaur research: observations and techniques

In this chapter, various lines of investigation are outlined as a way of reinforcing the point that it is often better to use a multiplicity of approaches if we are ever to gain a balanced understanding of the natural history and biology of fossil animals. But we are governed by two factors: what we can observe through new discoveries, and the way in which *technical innovations* can enhance our ability to understand and interpret such discoveries. I'll now highlight a few examples from both or these areas.

Observing footprints and tracks

Some dinosaur research has a distinctly sleuth-like quality, perhaps none more so than *ichnology*—the study of footprints.

> There is no branch of detective science which is so important and so much neglected as the art of tracing footsteps. (A. Conan Doyle, *A Study in Scarlet*, 1887, ch. 14)

The study of dinosaur footprints has a surprisingly long history. Some of the first to be collected and exhibited were found in 1802 in Massachusetts by Pliny Moody while ploughing a field. These large three-toed prints were illustrated and described by Edward Hitchcock in 1836 as the tracks of gigantic birds; some can still be seen today at the Beneski Museum of Natural History at Amherst

College, Massachusetts, while some of Hitchcock's specimens are preserved at the Sedgwick Museum, Cambridge (UK). From the mid-19th century onwards, individual prints or even trackways were discovered in various parts of the world. With the development of an understanding of the anatomy of dinosaurs, and more particularly the shape of their feet, it was realized that such large, apparently 'bird-like', three-toed prints that were found in Mesozoic rocks were made by dinosaurs rather than gigantic birds. Such tracks, though of local interest, were rarely regarded as of great scientific value. However, prompted by Martin Lockley of the University of Colorado at Denver, it is now appreciated that footprints (as Sherlock Holmes knew!) can yield a great deal of information.

Obviously footprints were left behind by *living* creatures (dinosaurs, in this instance) and such prints record the shape of the foot and the number of toes, which can often help to narrow down the likely track maker (especially if dinosaur skeletons have been discovered in similarly aged rocks nearby). While individual prints are intrinsically interesting, a set of tracks is particularly so, providing a record of the creature moving. They reveal the pose of each foot as it contacted the ground, the length of the stride, and the width of the track (how closely the right and left feet were spaced); using this evidence, it is possible to reconstruct how the legs moved in a mechanical sense. Furthermore, taking observations using data from a wide range of living animals it has also proved possible to calculate the speeds at which animals leaving the tracks were moving. These estimates are arrived at by simply measuring the size of the prints and the length of each stride, and making an estimation of the length of the leg. Although the latter might seem at first sight difficult to do with great accuracy, the actual size of the footprints has proved to be a remarkably good guide (judged from measurements of currently living animals), and of course skeletons of dinosaurs that lived at the time the tracks were made are sometimes already known.

The overall shape of individual prints may also reveal information about how such animals were moving: relatively flat, broad prints suggest that the whole foot was in contact with the ground for quite a long time, and that the track maker was moving relatively slowly; in other instances, the prints may just reveal the tips of the toes touching the ground—suggestive of an animal sprinting on the tips of its toes.

Another interesting aspect of dinosaur tracks relates to the circumstances that led to them being preserved at all. Footprints can never be preserved on a hard surface, it needs to be soft and slightly moist (ideally of a very thick pasty consistency). Once prints have been made, it is then important that they are not greatly disturbed before the soil solidifies; this can happen if the prints are buried quickly beneath another layer of mud, provided the imprinted surface has become baked hard in the sun. Alternatively in hot environments high rates of evaporation allow minerals to be precipitated within the fabric of the soil to form a type of 'cement'. Quite often it is possible to deduce, from details of the sediment in which the tracks were made, exactly what the environmental conditions were like when the dinosaur left its tracks. This can range from the degree to which the mud was disturbed by the feet of the animal and how deeply the feet sank into the sediment, to how the sediment seems to have responded to the movements of the foot. Sometimes it can be seen that a creature was moving up or down slopes simply from the way sediment is scuffed up in front of, or behind, each print. Tracks left by dinosaurs can therefore offer a great deal of information about how they moved and the types of environments in which they moved.

The study of tracks can also reveal information about dinosaur social behaviour. Multiple tracks of dinosaurs, such as one discovered in the Paluxy River at Glen Rose in Texas, revealed two parallel tracks, one left by a huge sauropod ('brontosaur'—see the sauropod silhouettes in Figure 29, later in the volume) and the other by a large carnivorous theropod dinosaur ('allosaur'; again, see Figure 29). The tracks seemed to show the allosaur tracks converging on those of the brontosaur. At the intersection of the tracks, one print is missing, and it was proposed that this indicated the point of attack. However, Lockley was able to show from detailed maps of the track site that the brontosaurs (there were several) continued walking beyond the supposed point of attack; and, even though the large allosaur was following the brontosaur (some of its prints overlap those of the brontosaur), there is no sign of a 'scuffle'. Very probably this predator was simply tracking potential prey animals but following at a 'safe' distance, perhaps waiting for a straggler to become detached from the main herd. Another set of trackways at Davenport Ranch, also in Texas displays tracks of twenty-three 'brontosaurs' walking in the same direction at the same time. This suggests that some dinosaurs moved in herds. Herding or gregarious behaviour is impossible to deduce from isolated skeletons, but trackways provide indisputable evidence.

Observing 'earthworks'

Some fascinating evidence, also from Colorado, may provide a side-light on mating rituals among dinosaurs. In 2015 a team led by Martin Lockley discovered an area covered with large, paired, scoop-shaped depressions. At first it seemed that these might have just been slightly odd sedimentary structures, but a similar pattern was found at a number of other locations. The large elongate depressions (approximately 2 metres by 1 metre) are flanked by mounds of displaced soil and the sides of each scoop are marked by deep scratches; it seems that a large animal with sharp-clawed feet had purposely excavated these structures: but why? The solution that the team came up with was that the areas where these odd structures are found most probably represent what in the bird-world of today are called *leks*: discrete areas of land ('stages') purposely constructed so male birds can perform

26. Dinolek restoration. Tyrannosaur-like theropods are shown performing on their 'leks' as part of their ritualized mating behaviour.

ritualized dances or displays to attract mates. Judged by their size and their geological age, these 'dinoleks' may well have been excavated by large tyrannosaur-like theropod dinosaurs. The thought that large theropods may have indulged in behaviours that we today associate exclusively with birds (Figure 26) is perhaps not so surprising given the close relationship between birds and theropods (Figure 29); however, if this interpretation is correct it might now be said, with some justification, that modern birds exhibit mating behaviours first established by dinosaurs in the Mesozoic.

Observing distorted rocks

Dinosaur tracks have sometimes been found in areas that have not yielded skeletal remains of dinosaurs, so tracks may help to fill in gaps in the fossil record of dinosaurs. Interesting geological concepts have also emerged from a consideration of dinosaur track properties. Some of the large sauropod dinosaurs (the 'brontosaurs' referred to earlier) may have weighed as much as 20 tonnes in life. Such animals would have exerted enormous forces on the ground when they walked. On soft substrate the pressure from the feet of such dinosaurs may have distorted the layers of soil to a depth of a metre or more beneath the surface—creating a series of 'underprints' of the surface footprint. If herds of such enormous creatures trampled over areas, as they certainly did at Davenport Ranch, then they also had the capacity to greatly disturb the soil beneath—pounding it up and destroying its normal sedimentary structure. This phenomenon has been named *dinoturbation*.

Dinoturbation might be just a sedimentary phenomenon, but it also hints at another biological impact of dinosaur activity that might eventually prove to be measurable over time: the ecological impact of dinosaurs on terrestrial plant communities. Herds of multi-tonne dinosaurs moving across a landscape may well have been capable of devastating the local ecology. We know that 3–4 tonne elephants can cause considerable damage to the African bush because of the way that they can literally tear up and knock down mature trees. What might a herd of 20 tonne 'brontosaurs' have done? Did this type of destructive activity have a measurable effect upon the other animals and plants living at the time? Did 'brontosaurs' play some role in the evolutionary history of plant and animal life during the Mesozoic? These are tantalizing questions—without any clear answer at present.

Observing dung

A peculiarly unromantic branch of palaeobiological investigation focuses on the dung of animals such as dinosaurs. *Coprolites* ('copros' = dung; 'lithos' = stone) and their study has a long history. The recognition of the importance of preserved dung dates back to the work of William Buckland of Oxford University (the man who scientifically described the first dinosaur, *Megalosaurus*). A pioneering geologist from the first half of the 19th century, Buckland spent considerable time collecting and studying rocks and fossils of his native area around Lyme Regis in Dorset. As he was doing so, Buckland noted large numbers of distinctive pebbles that often had a faint spiral shape. On closer inspection, breaking them open with a hammer as well as looking at thin, polished sections using a microscope, Buckland was able to identify shiny fish scales, pieces of bone, and the sharp hooks from *belemnite* (an ancient squid-like mollusc) tentacles. He concluded that these curious stones were most probably the petrified excreta of the predatory sea reptiles also found in these rocks. Although considered somewhat 'distasteful' at the time, given the Regency obsession with 'sensibilities', the study of coprolites offered direct evidence of the diet of very ancient creatures. It may be noted that Buckland had an impish sense of humour. After he had made his discovery of the origin of such stones (known locally as 'bezoars') he had many of them cut in half, polished, and inset into small 'bezoar tables' for genteel ladies to drink tea from—little did they know what those 'pretty polished stones' really were!

As was the case with footprints, the question, 'Who left this behind?', though obviously amusing, can present significant problems. Occasionally, coprolites, or indeed gut contents (coiled *cololites*—many of Buckland's 'bezoars' were, in fact, cololites), have been discovered preserved within the bodies of some fossil vertebrates (notably fish); however, it has proved particularly difficult to connect coprolite fossils to specific dinosaurs or even groups of dinosaurs. Karen Chin of the US Geological Survey has devoted herself to the study of coprolites and had singular difficulty in reliably identifying dinosaur coprolites—until....

In 1998, Chin and colleagues were able to report the discovery of what they referred to in the title of their article as 'A *king*-sized theropod coprolite'. The specimen in question was discovered in Maastrichtian (latest Cretaceous) sediments in Saskatchewan and comprised a rather knobbly lump of material, about 40 cm long and with a volume of approximately 2.5 litres. Immediately around and inside the specimen were broken fragments of bone and a finer, sand-like, bone powder. Chemical analysis revealed high levels of calcium and phosphorus, confirming a high concentration of the bone mineral *apatite* (calcium phosphate). Histological thin-sections of the fragments further confirmed the cellular structure of bone and some showed they belonged to a small ceratopian dinosaur. The specimen was a coprolite from a large carnivore. Surveying the fauna known from the rocks in this area, the only creature sufficiently large to have been able to pass dung of these dimensions (remember this specimen was 'merely' the dried remnant of the original dung; it would have had a larger volume when freshly deposited) was the large theropod *Tyrannosaurus rex* (*T. rex*; 'King of the tyrant reptiles'). The fact that not all the ceratopian bones had been digested indicated that the food had moved through the gut very speedily, suggesting *T. rex* was a hungry endotherm.

Observing illnesses

The confirmation of a diet of meat in *T. rex* is clearly not exactly unexpected, given its overall anatomy. However, an interesting pathological consequence of a diet rich in meat has also been detected within the skeleton of *Tyrannosaurus*.

'Sue', the nickname of the large skeleton of *Tyrannosaurus rex*
now on display at the Field Museum in Chicago, is interesting
because it shows a range of pathologies. One of its hand bones
(*metacarpals*) exhibits smoothly rounded pits at the joint with its
first finger bone (*phalanx*); these pits were examined by
pathologists as well as palaeontologists. The palaeontologists
discovered that other tyrannosaurs exhibit similar lesions. The
pathologist was able to confirm, following detailed comparison
with pathologies from living reptiles and birds, that the lesions
were mostly likely caused by gout. This illness, also known in
humans, generally affects the feet and hands, and is extremely
painful—causing swelling and inflammation around the joints,
caused by the deposition of abrasive urate crystals. In humans,
one factor linked to gout is a diet rich in purine, which is found
abundantly in red meat. So, *Tyrannosaurus* not only looked like a
meat-eater, its dung proves it and so does one of its diseases.

'Sue' also displays a large number of tell-tale remains of past
injuries. When bones are broken during life, they have the
capacity to heal themselves. Although modern surgical techniques
enable repair of broken bones with considerable precision, in
Nature the broken ends of the bone do not usually align
themselves precisely and a thick *callus* forms around the area
where the ends of bone meet. Such imperfections in the repair
process leave marks on the skeleton that can be detected after
death. It is clear that 'Sue' suffered a number of injuries. On one
occasion, she experienced a major trauma to the chest (several
broken and re-healed ribs). In addition, her back and tail show a
number of breakages along her spine.

The surprising aspect of these observations is that an animal such
as *T. rex* was able to survive periods of injury and sickness.
It might be predicted that a large predator such as *T. rex* would
become extremely vulnerable, and therefore potential prey itself,
once it was injured. That this did not happen (at least in the
instance of 'Sue') suggests either that such animals were

extraordinarily tough and not unduly affected by quite serious trauma, or that these dinosaurs may have lived in socially cohesive groups that might have acted co-operatively to assist an injured individual.

Other pathologies have also been noted in various dinosaurs. These range from destructive bone lesions resulting from periodontal abscesses (in the case of jaw bones), or septic arthritis and chronic osteomyelitis in other parts of the skull or skeleton. One particularly unpleasant example of long-term infection of a leg wound was recorded in a small ornithopod. The partial skeleton of this animal was discovered in Early Cretaceous sediments in south-eastern Australia. The hindlimbs and pelvis were well preserved, but the lower part of the left leg was grossly distorted and shortened. Although the actual cause of the injury could not be ascertained, it was suspected that the animal was bitten on the shin, close to the knee, of its left leg. As a result, the fossilized bones of the shin (tibia and fibula) had become severely over-grown by a huge, irregular mass of bone.

Examination and X-radiography of the fossil shin bone shows that instead of remaining localized, an infection spread down its marrow cavity, partially destroying the bone as it went. As the infection spread, extra bony tissue was added to the exterior of the bone as if the body was trying to create its own 'splint'. It appears that the animal's immune system was unable to prevent the spread of infection and large abscesses formed beneath the outer bony sheath; the pus from these must have leaked through from the leg bones and may have run out on to the surface of the skin as a stinking, weeping sore. Judged from the amount of bone growth around the injury, this animal lived for several months, while suffering this horribly crippling injury, before it finally succumbed.

Tumours have rarely been recognized in dinosaur bones. The most obvious drawback with trying to study the frequency of cancers in dinosaurs has been the need to destroy dinosaurian bone in order to make histological sections—obviously something that has little appeal to museum curators. Recently, Bruce Rothschild developed a technique for scanning dinosaur bones using X-rays and fluoroscopy. The technique is limited to bones less than 28 cm in diameter, and for this reason he surveyed large numbers (over 10,000) of dinosaur vertebrae. The vertebrae came from representatives of all the major dinosaur groups. He discovered that cancerous tumours were very rare and mainly restricted to hadrosaurian ornithopods.

As recently as 2016, colleagues working on some newly discovered skull bones of the hadrosaur (*Telmatosaurus*) from Romania, noted that one of the jaws was deformed. CT (computed tomography) scanning revealed that the deformity was caused by the growth of a benign tumour (an ameloblastoma) within the jaw itself. While this dinosaur's jaw would have been swollen, the tumour would not itself have been fatal; we humans also suffer from this sort of benign tumour growth today.

Quite why tumours should be largely restricted to hadrosaurs is puzzling. Rothschild did wonder whether the diets of hadrosaurs had a bearing on this epidemiology. Rare discoveries of 'mummified' carcasses of hadrosaurs reveal quantities of conifer tissue; today conifers are known to contain high concentrations of tumour-inducing (alkaloid) chemicals. Whether it is a case of genetic predisposition to cancer among hadrosaurs or, as Rothschild suggests, an environmental factor (a mutagenic diet) is mere speculation.

Technical approaches: how dinosaurs moved

While footprints give us vivid evidence of dinosaurs standing and using their feet when walking or running, and occasionally yielding insights into behaviour, they are very limited with respect to what such animals *could* do. Trackways give an indication of speed of movement, but don't animals just amble around most of the time? I certainly do, even though I am capable of running moderately fast—and watching lions or even cheetahs in the wild reveals that an awful lot of their time is spent resting and walking slowly; very rarely do they run at top speed. With this in mind the skeletons of dinosaurs can also be examined more closely in order to determine their *maximum potential running speed*. How often has it been asked whether *T. rex* could really run as fast as an off-road vehicle since the screening of *Jurassic Park* and its sequels?

Limb proportions are a good preliminary guide to an animal's potential to move at speed, or more ponderously; these two extremes are known as *cursorial* (~fast) and *graviportal* (~slow). *Cursorial* animals tend to have light bodies and their legs are slender with short thighs and longer shins and toes; this combination means that the limb can be moved quickly by muscles bunched close to the hips because it is lighter toward the toes and can take long strides—it's the same reason that we run wearing light shoes rather than heavy boots! *Graviportal* animals have long thighs, shorter shins and stubby, spreading feet and toes—their legs move more slowly, the leg muscles create more power, and the feet are designed to spread and distribute weight more evenly: Nature designs them with the equivalent of heavy walking boots.

Taking limb measurements a little further it is possible to compare the limb proportions of dinosaurs to those seen in living animals whose maximum speeds are known. If these observations are taken in tandem with studies of the histology of dinosaur bone it is possible to estimate the overall strength of dinosaur limb bones and their ability to resist the bending stresses that would have been imposed upon them during a step cycle. Perhaps not surprisingly small theropods (dromaeosaurs and their kin) have cursorial limb proportions and seem to be fast runners, whereas their much larger cousins (tyrannosaurs) had lower bone strength and limb proportion indicators and would *not* have been capable of very fast running (which will disappoint *Jurassic Park* fans). Equally the large and obviously heavy dinosaurs— sauropodomorphs, stegosaurs, ankylosaurs, and ceratopians (Figure 29) all display graviportal limb proportions and comparatively weak limb bones that were poorly adapted for dealing with large bending stresses; these latter animals swung their legs like pendulums underneath their bodies, rather than flexing their knee and ankle joints as do cursors.

Bioengineering-style approaches have also begun to be used in order to estimate: the maximum speed that might have been attained by dinosaurs; the range of potential limb postures that could have been adopted; the size, length, and strength of the various muscles that moved the limbs and in some instances the manoeuvrability and flexibility of dinosaurs, such as their ability to change direction quickly, their 'turning-circles', or even their ability to squat to lay eggs on the ground without breaking them. As a research area the bioengineering approach represents a shift toward greater precision in our understanding of how such long-extinct animals moved: their strength, their power, their endurance, and their agility. Leading workers in this area of research include Steve Gatesy (Brown University), John Hutchinson (Royal Veterinary College, London), and Don Henderson (Tyrrell Museum, Alberta). Earlier approaches were, in effect, simply saying: 'well these could move quite quickly and these others rather more slowly'; nowadays, attempts are being made to quantify how quickly or how slowly, and to investigate the links between physiology and locomotor abilities with the intention of discovering greater scientific accuracy.

Technical approaches: isotope geochemistry

Geochemistry uses radioactive isotopes of oxygen, particularly oxygen-16 and oxygen-18, and their proportions in chemicals (carbonates) found in the shells of microscopic marine organisms, to estimate the temperature of ancient oceans, and global climatic conditions. Basically, the understanding is that the higher the proportion of oxygen-18 (compared to oxygen-16) locked into the chemicals within the shells of these organisms, the colder the temperature of the ocean in which the organisms originally lived.

In the early 1990s, a palaeontologist (Reese Barrick) and a geochemist (William Showers) collaborated to see if it might be possible to do something similar with the chemicals in bones—particularly the oxygen that forms part of the phosphate molecule in bone minerals. They first applied this approach to some known vertebrates (cows and lizards) by taking samples of bones from different parts of the body (ribs, legs, and tail) and measured the oxygen isotope proportions. Their results showed that for the endothermic mammal (cow) there was very little difference in the body temperature between the bones of the legs and ribs; as might be expected, the animal has a constant body temperature. In the lizard, however, the tail was between 2°C and 9°C lower than its ribs; the ectotherm did not have such an even distribution of body heat: the peripheral parts were, on average, cooler than the body core.

Barrick and Showers then performed a similar analysis on various bones from a well-preserved *T. rex* skeleton collected in Montana. Drilled samples from ribs, leg, toe, and tail bones revealed a mammal-like result: the oxygen isotope ratios differed very little, indicating that the body had a fairly even temperature throughout. This supported the idea that dinosaurs had a very even body temperature. More recent work by these authors seems to confirm these basic findings and extended this pattern to a range of other dinosaurs, including hadrosaurs.

As is often the case, such results generated lively discussion. There were concerns that the bones may have been chemically altered during fossilization, which would render the isotopic signals meaningless; whereas physiologically minded palaeobiologists were far from convinced about what the result meant: an even temperature signal is entirely consistent with the idea that most dinosaurs were large-bodied mass-homeotherms (Chapter 5) but gave no conclusive evidence of endothermy.

Technical approaches: the scanning revolution

The steady improvement in technology and its potential to be used to explore palaeobiological questions, has become widespread in recent years. A few of these approaches will be examined in the next section. They are not without their limitations and pitfalls, but in some instances questions may now be asked that could not have been dreamt of ten years ago.

One of the most anguished dilemmas faced by palaeobiologists is the desire to explore as much of any new fossil as possible (using mechanical tools) but at the same time to minimize the damage caused to the specimen by such action. The discovery of the potential for X-rays to create images on photographic film of the interior of the body has been of enormous importance to medical science. The more recent revolution in medical imaging, through the development of CT and MRI (magnetic resonance imaging) scanning techniques that are linked directly to powerful data-processing computers, has resulted in the ability to create three-dimensional images that allow researchers to see inside objects such as the human body or other complex structures that would only normally be possible after major exploratory surgery.

The potential to use CT scanning to see inside fossils was quickly appreciated. It began, curiously enough, through contacts with aero-engineers because they were interested in very large CT scanners capable of examining aero-engines for evidence of tiny cracks in turbo-fans or other potential points of failure. Such large, high-powered scanners had the potential to see into rocks—provided that the engineers didn't mind too much if we borrowed their machines. Gradually, as the cost of more powerful scanners decreased they became more readily affordable to palaeontologists.

One of the pioneers in this field is Tim Rowe, along with his team based at the University of Texas in Austin. He has managed to set up a fossil-dedicated, high-resolution CT scanning system, and, as we shall see, this and other resources have been put to some extremely interesting uses. Larry Witmer has done something rather similar at the University of Ohio (see Figure 27).

Honking hadrosaurs

One use of CT scanning can be demonstrated using the example of hadrosaurian ornithopods. These dinosaurs were abundant in Late Cretaceous times and have remarkably similarly shaped bodies; they only really differ in the shape of their headgear, but the reason for this difference has been a long-standing puzzle. When the first crested dinosaur was described in 1914, it was considered that these were simply interesting decorative features. However, in 1920 it was discovered that these crests were composed of thin sheaths of bone that enclosed tubular cavities or chambers of considerable complexity.

Theories to explain the purpose of these cavernous crests abounded from the 1920s onwards. The very earliest claimed that the crest provided an attachment area for ligaments running from the shoulders to the neck that supported the large and heavy head. From then on, ideas ranged from their use as weapons; that they carried highly developed organs of smell; that they were sexually specific (males had crests and females did not); and, the most far-sighted, that the chambers might have served as vocal resonators, as in modern birds. During the 1940s, there was a preference for aquatic theories: it was suggested they either formed a reserve air-tank when swimming underwater, or that the crests created an air-lock to prevent water flooding the lungs when these animals fed on underwater weeds.

Most of the more outlandish suggestions have been abandoned, either because they were physically impossible or they do not

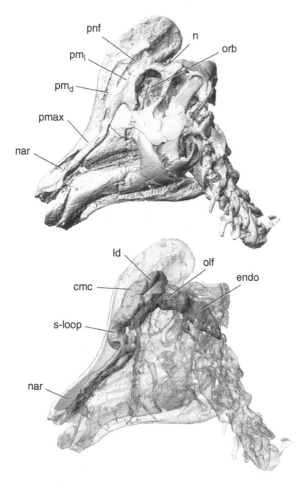

27. Computed Tomography (CT) scanned image of the surface of the
skull and top of the neck of a hadrosaur dinosaur, *Corythosaurus*. The
skull is comparatively little deformed and the enlarged helmet-like
crest is clearly visible (top). The CT scan image (top) has been
manipulated digitally so that the surface bones are rendered
semi-transparent (as if made of clear perspex) so that the internal
cavities within the crest can be visualized, without the need to break
open and destroy such a beautiful specimen. The convoluted tubes of
the nasal passages can be clearly seen and these connect at the rear to
the brain cavity (bottom).

accord with known anatomy. What is now thought is that the crests probably performed a number of interrelated functions of a mainly social/sexual type. Their distinctive shape undoubtedly provided a visual social recognition system for individual species: if we can see that they are clearly different then no doubt the living dinosaurs could as well! And, of course the crests would have naturally served as sexual display structures, to help them attract mates. As can be seen in Figure 27 the cavities were complex and could have housed a well-developed sensory area, giving them an acute sense of smell. Some styles of hadrosaur crest were sufficiently robust to have been used either in flank or head-butting activities as part of pre-mating rituals or male–male rivalry competitions.

Finally, the expanded chambers and tubular areas, associated with either the crests or the snout, are thought to have functioned as resonators. Again, this presumed vocal ability (seen today in birds and also crocodiles) can be linked to aspects of social behaviour in these dinosaurs. A colleague (David Weishampel, at Johns Hopkins University) studied the tubular crest of the dinosaur *Parasaurolophus* (see the hadrosaur silhouette in Figure 29). He made a crude model of the internal crest using simple plastic pipes and U-bends (from plumbing kits!) and, by adding a mouthpiece, he was able to blow into his model and it produced a deep, foghorn-like sound.

One of the greatest problems associated with the resonator theory was gaining direct access to skull material that would allow detailed reconstruction of the internal air passages, without breaking open prized and delicate specimens. CT techniques (Figure 27) made such internal investigations possible. Crests of *Parasaurolophus* that were sufficiently well preserved have now

Abbreviations: nar—bony margin of the nostril; n—nasal bone; orb—cavity for the eyeball; pmax—premaxilla, which forms the majority of the crest structure (including pmd, pml, and pnf); s-loop, cmc, and ld—refer to parts of the nasal passages; olf—represents the olfactory area of the nasal passages (sense of smell); endo—refers to the cavity within which the brain and its associated structures (pituitary body, inner ear, cranial nerves, and blood vessels) are located.

been CT scanned and the digital images processed so that the spaces *inside* the crest, rather than the bone of the crest, can be visualized. The rendered image of the crest's interior revealed an extraordinary degree of complexity (far greater than David Weishampel's 'trombone'). Several parallel, narrow tubes looped tightly within the crest, creating the equivalent of an orchestral cluster of trombones. There is little doubt that the crest cavities in animals like *Parasaurolophus* acted as resonators and that the organ-like piping could probably produce a more harmonic range of foghorn-like tones (...perhaps even a melody!).

Technical approaches: how dinosaurs fed

An innovative way of using CT imaging was developed by my former research student Emily Rayfield and colleagues, first of all at the University of Cambridge and then later at Bristol University. Using CT images, sophisticated computer software, and a great deal of biological and palaeobiological information, it proved possible to investigate how the skulls of dinosaurs may have functioned mechanically when they were alive.

We know, in very general terms, that *Allosaurus* (Figures 28 and 29) was a predatory creature that probably fed on a range of prey living in Late Jurassic times. Sometimes tooth marks or scratches have been found on fossil bones and these can be matched up against the teeth in the jaws of an allosaur as a form of 'proof' of the guilty party. But what does such evidence tell us? The answer is: not as much as we might like. We cannot be sure if the tooth marks were left by a scavenger feeding on a carcass or by the actual killer. Equally, we cannot tell what style of predator an allosaur might have been: did it run down its prey after a long chase, or did it lurk and pounce? Did it have a devastating bone-crushing bite, or was it more of a cut and slasher?

Rayfield was able to obtain CT scan data created from an exceptionally well-preserved skull of the Late Jurassic theropod

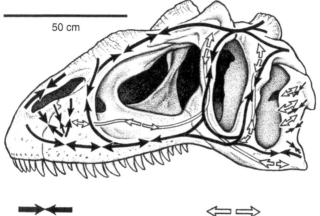

Compression   Tension

28. *Allosaurus*: showing the lines of stress and strain that are induced in the skull bones when the virtual skull is modelled biting food. Simplified stress-strain model based on analysis of FEA strain patterns.

Allosaurus. High-resolution scans of the skull were used to create a very detailed three-dimensional image of the entire skull. However, rather than simply creating a beautiful hologram-like representation of the skull, Rayfield converted the image data into a three-dimensional 'mesh'. The mesh consisted of a series of point coordinates (rather like the coordinates on a topographic map), each point was linked to its immediate neighbours by short 'elements'. This created what in engineering terms is known as a *finite element map* of the entire skull: nothing as complicated as this had ever been attempted before.

The remarkable property of this type of model is that with the appropriate computer and software it is possible to record, on the finite element map, the material properties of the skull bones, for example the strength of skull bone, of tooth enamel, or of cartilage on the joints between bones. In this way, each 'element' can be prompted to behave as though it were a piece of real skull, and each element is linked to its neighbours as an integrated unit, as it would have been in life.

Having mapped the virtual skull of this dinosaur in finite elements, it was then necessary to work out how powerful its jaw muscles were in life. Using modelling clay, Rayfield was able to quite literally shape and mould the jaw muscles of this dinosaur. Once she had done this, she was able to calculate from their dimensions—their length, girth, and angle of attachment to the jaw bones—the amount of force that they could have generated. To ensure that these calculations were as realistic as possible, two sets of force estimates were generated: one based on the view that dinosaurs like this one had a rather crocodile-like (ectotherm) physiology, the other assumed an avian/mammalian (endotherm) physiology.

Using these sets of data, it was then possible to superimpose these forces on the finite element model of the *Allosaurus* skull and quite literally 'test' how the skull would respond to maximum bite forces, and how these would be distributed (as lines of stress and areas of strain—Figure 28) within the skull. The experiments were intended to probe the construction and shape of the skull, and the way it responded to the stresses imposed upon it during feeding.

What emerged was fascinating. The skull was extraordinarily strong (despite all the large holes over its surface that might be thought to have weakened it significantly). In fact, the holes proved to be an important part of the strength of the skull. When the virtual skull was tested until it began to 'yield' (that is to say, it was subjected to forces that were beginning to fracture its bones), it was found to be capable of withstanding up to twenty-four times the force that the jaw muscles could exert when they were biting as hard as 'allosaurianly' possible.

What became obvious from this work was that the allosaur skull was hugely over-engineered in relation to the power that could be exerted by its jaw muscles. Natural selection usually provides a 'safety factor' in the design of most skeletal features: a sort of trade-off between the amount of energy and materials needed to build that part of the skeleton and its overall strength under normal conditions of life. That 'safety factor' varies, but is generally in the range of two to five times the forces experienced during normal life activities. To have the skull of *Allosaurus* built with a 'safety factor' of twenty-four seemed ludicrous.

Re-examination of the skull, and a rethink about its potential methods of feeding, led to the following realization: the lower jaw was physically quite slender and 'weak' in the way it was constructed, so the animal probably did have a comparatively low-power bite, particularly with respect to the overall skull strength. Perhaps, we reasoned, the skull was constructed to withstand very large forces (in excess of 5 tonnes) for other reasons. The most obvious was that the skull may have been used as the principal attack weapon—rather like an axe. These animals may well have lunged at their prey with the jaws opened very wide, and then slammed their head downward against their prey in a potentially devastating, slashing blow. With the weight of the body behind this movement, and the resistance of the prey animal, the skull would need to be capable of withstanding extremely high short-term loads.

Once the prey had been subdued following the first attack, the jaws could then be used to bite off pieces of flesh in the conventional way, but this might reasonably have been aided by using the legs and body to assist with tugs at resistant pieces of meat, again loading the skull quite highly through forces generated by the neck, back, and leg muscles.

In this analysis, it has proved possible to gain an idea of *how* predation and feeding may have been achieved in allosaurs in ways that until a few years ago would have been unimaginable. Yet again, the interplay between new technologies and different branches of science (in this instance engineering design and new computational methods) have enabled us to probe palaeobiological problems and generate new and interesting observations. We still cannot be sure whether *Allosaurus* was a pursuit predator or a pouncer, but given what we know now, my bet would be on the latter.

Technical approaches: studying biomolecules and tissues

I cannot finish this chapter without mentioning the *Jurassic Park* scenario: discovering dinosaur DNA, using modern biotechnology to reconstitute that DNA, and using this to bring a dinosaur back to life.

There have been sporadic reports of finding fragments of dinosaur DNA in the scientific literature over the past decade, and then using *PCR* (polymerase chain reaction) biotechnology to amplify the fragments so that they can be studied more easily. Unfortunately, for those who wish to believe in the Hollywood-style scenario, absolutely none of these reports have been verified, and in truth it is exceedingly unlikely that any genuine dinosaur DNA will ever be isolated from dinosaur bone. It is simply the case that DNA is a long and complex biomolecule which degrades over time in the absence of the metabolic machinery that will maintain and repair it, as happens in living cells. The chances of any such material surviving unaltered for over sixty-six million years while it is buried in the ground (and subject there to all the contamination risks presented by micro-organisms and other biological and chemical sources, as well as ground water) are effectively zero.

All reports of dino-DNA to date have proved, upon further analysis, to be records of contaminants. In fact the only reliable fossil DNA that has been identified is far more recent, and even these discoveries have been made possible because of unusual preservational conditions. For example, brown bear fossils whose remains are dated back to about 60,000 years have yielded short strings of mitochondrial DNA—but these fossils had been frozen in permafrost since the animals died, providing the best chance of reducing the rate of degradation of these molecules. Dinosaur remains are of course 1,000 times more ancient than those of arctic brown bears. Although it might prove possible to identify some dinosaur-like genes in the DNA of living birds, regenerating an entire dinosaur is beyond the bounds of science.

One extremely interesting set of observations concerns the analysis of the appearance and chemical composition of the interior of some tyrannosaur bones from Montana. Mary Schweitzer and colleagues from North Carolina State University were given access to a remarkably well-preserved young *T. rex* skeleton collected by Jack Horner (the real-life 'Dr Alan Grant' in the film *Jurassic Park*). Examination of the skeleton suggested that there had been very little alteration of the internal structure of its bones; indeed, so unaltered were they that the individual bones of the tyrannosaur had a density that was consistent with that of modern animal bones that had simply been left to decay and then dry.

Schweitzer was looking for ancient biomolecules, or at least any remnant chemical signals that they might have left behind. Having extracted material from the interior of the bones, this was powdered and subjected to a broad range of physical, chemical, and biological analyses. The idea behind this approach was not only to have the best chance of 'catching' some trace, but also to have a range of semi-independent support for the signal, if it emerged. The burden really is upon the researcher to find some positive proof of the presence of such biomolecules; the time elapsed since death and burial, and the overwhelming probability that any remnant of such molecules has been completely destroyed or flushed away, seem to be overwhelming. Nuclear magnetic resonance and electron spin resonance revealed the presence of molecular residues resembling *haemoglobin* (the primary chemical constituent of red blood cells); spectroscopic analysis and *HPLC* (high performance liquid chromatography) generated data that was also consistent with the presence of remnants of the haeme molecule. Finally, the dinosaur bone tissue was flushed with solvents to extract any remaining organic residue. This extract was then injected into laboratory rats to see if it would raise an immune response—and it did! The antiserum created by the rats reacted positively with purified avian and mammalian haemoglobins. From this set of analyses, it seems very probable that chemical remnants of dinosaurian haemoglobin compounds were preserved within these *T. rex* bones.

Even more tantalizingly, when thin sections of portions of bone were examined microscopically, small, rounded microstructures could be identified in the vascular channels (blood vessels) within the bone. These microstructures were analysed and found to be notably iron-rich compared to the surrounding tissues (iron being a principal constituent of the haeme molecule). Also the size and general appearance was remarkably reminiscent of avian nucleated blood cells. Although these structures are not actual blood cells, they certainly seem to be the chemically altered 'ghosts' of the originals. Quite how these structures have survived in this state for at least 67 Ma is a considerable puzzle.

Schweitzer and her co-workers have also been able to identify (using immunological techniques similar to the one mentioned earlier) biomolecular remnants of the 'tough' proteins known as collagen (a major constituent of natural bone, as well as ligaments and tendons) and keratin (the material that forms scales, feathers, hair, and claws).

Although these results have been treated with considerable scepticism by the research community at large—and rightly so, for the reasons elaborated above—nevertheless, the range of scientific methodologies employed to support their conclusions, and the exemplary caution with which these observations were announced, represent a model of clarity and application of scientific methodologies in this field of palaeobiology.

Chapter 7
The future of research on the past

The overall picture: dinosaurs in their world

There is an unmistakable grandeur about a Mesozoic World filled with a bewildering range and variety of dinosaurs. Many of them were huge, but they also explored a vast range of bodily forms and as a group (if we exclude, rather artificially, the birds that are without doubt living theropod dinosaurs) then this diversity dominated the Earth for over 170 million years. As a palaeontologist one inevitably becomes somewhat glib about numbers and the time that has passed, but just pause and reflect for a moment…ONE HUNDRED AND SEVENTY MILLION years! Just to put it into context (a little unfairly perhaps) we humans have a history that can be traced back through fossil remains a mere 500,000 years, a barely noticeable amount of time in the context of dinosaur world—and yet we do tend to think of ourselves as long-established on Earth.

Dinosaurs first appeared in a world that consisted of a single giant supercontinent named Pangea (Figure 23(a)) and could have walked across the entire world because there were no seaways to prevent them. As time passed the continents gradually broke apart (Figures 23(b) and (c)) to eventually form the geographic areas that we recognize today (Figure 23(d)). Seas separated these newly formed continents and this changed profoundly the climatic patterns on Earth—the world became hot and humid, rather than hot and very arid, as it had been when the dinosaurs first made their appearance.

That is a very crude geophysical backdrop to dinosaur world, but what of the dinosaurs themselves? How did these great changes affect them? How did they evolve over time? And, do they show distinct patterns in their evolutionary history? And, of course, that perennial question: why did all the non-avian dinosaurs become extinct sixty-six million years ago?

Patterns and processes. A considerable amount of effort is now being expended on trying to determine patterns of change across the vast expanse of time when the non-avian (and avian-theropod) dinosaurs are known to have been dominant on land. The fossil record of dinosaurs though rather sparse (we can only have discovered a small fraction of the actual number of species that existed during 170 million years of Earth history) is being explored quantitatively. Databases representing the number of known different dinosaurs collected to date have been painstakingly assembled so that estimates may be made of the range and variety of dinosaurs that existed at any one time so that these can be compared with other time periods. Family trees (cladograms) that have been constructed (Chapter 4) can be *calibrated*—that is to say the time-ranges during which each dinosaur grouping (clade) is known to have existed (from fossil discoveries) are plotted against the geological timescale (see Figures 19 and 29). This information can be used to see whether the time of appearance of the different clades is in the order predicted by the branching events in the cladogram, or whether there are incongruities. For example, dinosaur groups might only be known much later than predicted from the shape and branch points in the family tree. This could mean that there is a 'ghost range': that is to say fossils representing particular clades of dinosaurs ought to be found in older rocks but we simply haven't found them yet; or the fossil record is correct and the clade mistakenly appears too early, implying that the cladogram contains errors. Such inconsistencies may seem frustrating but are always a useful way of directing research into the fossil record for new material, as well as for double-checking and revising cladogram construction. What can emerge from such work are patterns of appearance that may have an underlying evolutionary explanation, so the overall *pattern* created can lead to a better, or more refined, understanding of the evolutionary *processes* that created that pattern. For example one of the dominant evolutionary trends revealed by a team led by former research student Roger Benson (Oxford University), is of increasing body size over time, as shown in Figure 29. But what is *really* interesting is that while most dinosaur clades tend to get bigger with the passage of time the one group that bucks that trend is the avian-theropods and their descendants the birds.

The creation of large databases that record information on the time of appearance and duration of species, their geographic location, their anatomical characters, and their dimensions can be researched in many different ways. Analysis of this data is a process that requires considerable time and patience (as well as a good understanding of the statistical algorithms that are now available to analyse data drawn out of the larger databases).

The axiom 'garbage in: garbage out' (*GIGO*—in computer-speak) must always be remembered. Databases often contain errors; this is not entirely unreasonable because humans input the data in the first place and 'to err is human' (Alexander Pope). In addition, the data may contain a variety of biases associated with the way the data was collected in the first place. Unless these can be adequately accounted for they risk having a pernicious influence, and may introduce errors into the results.

As shown in Figure 30, research led by former research student Richard Butler (Birmingham University) demonstrated an underlying correspondence between the numbers of dinosaurs collected at any one time interval and the availability of rocks

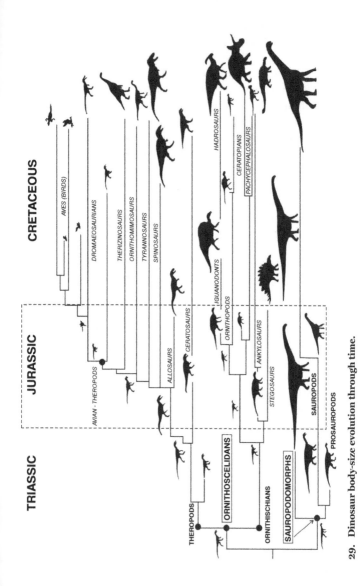

29. Dinosaur body-size evolution through time.

from which they have been collected. The graphical lines more or less match. This suggests that when comparing diversity patterns across time zones researchers need to account for the rock volume bias. Just because there appear to be more dinosaurs collected at one time it cannot simply be assumed that they were undergoing some evolutionary 'explosion'; it may simply be the case that more dinosaurs could be collected because there were more rocks available for collectors to explore.

Some of the most actively researched patterns in the fossil record are those concerning taxonomic diversity, morphological disparity, changes in faunal abundance across time, and the rates at which anatomical characters change across time. All of these approaches tend toward developing a better understanding of how and why dinosaurs originated when they did, how they evolved throughout their reign on Earth, and, ultimately, why the non-avian forms all succumbed so abruptly sixty-six million years ago.

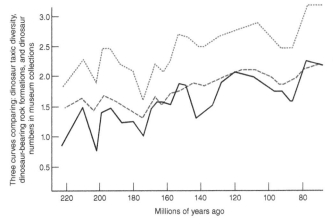

30. Bias estimation in fossil record. The solid black line represents a calibrated estimate of the changes in dinosaur diversity across the Mesozoic Era. The dashed line represents an estimation of the actual number of geological formations from which dinosaurs could be collected on Earth. The dotted line represents the number of dinosaur collections that have been made to date.

K-Pg extinctions: the end of dinosaurs?

Since the early decades of the 19th century, it had been known that different groups of organisms dominate different periods of Earth history. One of the more notable groups was the dinosaurs, and there was a steady reinforcement, from palaeontological surveys, of the idea that none was to be found in rocks younger than the end of the Cretaceous Period (approximately 66 Ma). In fact, it came to be recognized that the very end of the Cretaceous Period (abbreviated to K), leading into the Paleogene (abbreviated to Pg) now referred to as the *K-Pg boundary*, marked a major time of change (Figure 31) known as a mass-extinction event. Many species became extinct and were replaced in the Early Paleogene by a diversity of new forms: the K-Pg boundary therefore seemed to represent a major punctuation in life on Earth. The types of species that became extinct at this time included the fabled dinosaurs on land, of which there were many different varieties by Late Cretaceous times; a multiplicity of sea creatures, ranging from giant marine reptiles (mosasaurs, plesiosaurs, and ichthyosaurs), to the hugely abundant ammonites, as well as a great range of chalky planktonic organisms; while in the air the flying reptiles (pterosaurs) and enantiornithine birds disappeared forever.

Clearly it was necessary to try to understand what might have caused such a dramatic loss of life and resetting of Earth history. The flip side of this general question was just as important: why did some creatures survive? After all, modern birds survived, so did mammals, and so did lizards and snakes, crocodiles, and tortoises (Figure 31), fish and a whole host of other sea creatures. Was it just a matter of bad luck, or some subtle global change? Up until 1980, quite a few of the theories that had been put forward (and there were very many) to explain the K-Pg extinctions and survivals ranged from the sublime to the ridiculous.

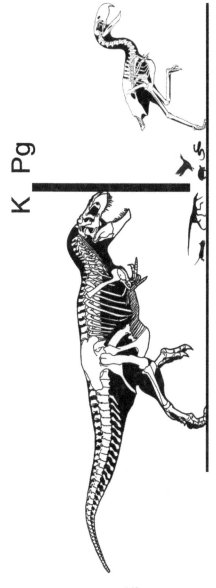

31. The K-Pg Extinction. At 66 Ma an event occurred that removed all non-avian dinosaurs from the Earth. This dramatic event is still puzzling, despite years of intensive research on the topic.

One of the more persistent of the pre-1980 theories (advocated
by Leigh Van Valen and Robert Sloan in particular) focused
upon detailed studies of the ecological composition of dinosaur
communities in the time zones closest to the K-Pg boundary.
The consensus from their work suggested that there was a shift
to progressively more seasonal/variable climatic conditions at
the end of the Cretaceous Period. This was mirrored in the
decline of those animals and plants less able to cope with
more stressful climatic conditions. This change was linked to
plate tectonics toward the close of the Cretaceous Period, which
could have been responsible for sea-level rise and increased
continental provinciality. The general impression was that the
world was gradually changing in character over the tens of
millions of years leading up to 66 Ma and, little by little, these
changes built into a crescendo represented by the dramatic
faunal and floral turnover.

Clearly such explanations require a long timescale for the
extinction event to take place, but the Achilles heel was that this
did not adequately account for the simultaneous changes seen in
marine communities. In the absence of better quality data,
arguments waxed and waned with no obvious resolution. It should
be noted that quite recently this general argument has been
re-visited by Sakamoto and colleagues following fresh censuses of
the abundance and diversity of dinosaur groups across the
geological Stages leading up to the extinction event 66 Ma. The
more detailed analyses echo the idea that dinosaurs were
declining in abundance and diversity before the actual time of the
mass-extinction, which is suggestive of some more subtle longer
term environmental effects depressing dinosaur abundance;
nevertheless there was still a rather sudden terminal event
representing a final 'coup-de-grace'.

Coming straight out of left field in 1980, decades of investigation
and disparate theories about the K-Pg extinction event were
completely overturned by, of all people the Nobel-Prize-winning

physicist/astronomer Luis Alvarez. His son Walter, a palaeobiologist, had been studying fine-resolution changes in plankton diversity in bands of clay that straddled the K-Pg boundary. One of the lines of thought at the time was that the actual geological interval between the Late Cretaceous and earliest Paleogene might simply represent a longish period of 'missing' time—a genuine gap in the continuity of the fossil record, so Walter was looking in detail at the clays that were formed at precisely that time interval. To assist Walter in his studies concerning the changes in planktonic communities across this critical time in Earth history, Luis suggested that he could measure the amount of time that had elapsed within the clays by sampling the cosmic dust that had accumulated in them. Cosmic dust falls continuously from the skies and brings with it the distinctive and measurable extra-terrestrial element, *Iridium*. Samples of clay were collected and Luis Alvarez had them analysed for their Iridium concentration. The results shocked both Luis and Walter Alvareze as well as the palaeontological and geological world. They found that the boundary layer, which was represented by a thin band of clay, contained anomalously high concentrations of Iridium. Not satisfied with the one result K-Pg clays were collected from elsewhere in the world: the same Iridium enrichment was seen. What emerged as a theory to explain such unprecedented levels of Iridium (and associated geological features such as 'shocked quartz' and tektite droplets) was that the only rational explanation had to be of a giant meteorite colliding with the Earth 66 Ma and that the collosal force of the impact would have vaporized the meteorite along with the local impact-site sediments.

They calculated that this meteorite would have needed to be at least 10 km in diameter. Considering the effect of the impact of such a giant meteorite, they further proposed that the huge debris cloud generated (containing water vapour and dust particles) after the impact would have shrouded the Earth completely for a significant period of time, perhaps several months or even a year or two. Shrouding the Earth in this way would have shut down photosynthesis of land plants and planktonic organisms, and could have triggered the collapse of terrestrial as well as aquatic ecosystems. At a stroke, the Alvarezes and their colleagues seemed to have found a unifying explanation for the K-Pg event.

As with all good theories, the new impact hypothesis generated an impressive volume of research. Throughout the 1980s, more and more teams of researchers were able to identify cosmic debris and violent impact-related signals in K-Pg boundary sediments from all four corners of the globe. By the late 1980s, the attention of a number of workers was drawn to the Caribbean area because the volume of impact-related debris appeared to be greatest in southern North America and the Caribbean islands. Reports showed that on some of the Caribbean islands, such as Haiti, deposits of sediments at the K-Pg boundary not only showed the impact Iridium signal, but immediately above this there was a great thickness of *breccia* (broken masses of rock that had been thrown together by huge tsunamis). This, as well as the greater thicknesses of the meteorite debris layer and its chemical signature, prompted the suggestion that the meteorite had impacted somewhere in the shallow sea in this area. In 1991, the announcement was made that researchers had identified a very large subterranean meteorite impact crater, which they called Chicxulub, on the Yucatán Peninsula of Mexico. The crater itself could not be seen on the land surface because it had been covered by sixty-six million years of sediment; it was visualized by studying seismic echoes of the Earth's crust (the equivalent of subterranean radar). The crater appeared to be approximately 200 km across and coincided with the K-Pg boundary layer: the Alvarez theory seemed to be independently vindicated in a most remarkable way.

From the early 1990s onwards, study of the K-Pg event shifted away from the causes, which then seemed to have been established, to attempting to link the extinctions at this time to a single catastrophic event.

The parallels to on-going 'nuclear winter' debates are fairly clear. Advances in computer modelling, combined with knowledge of the likely chemical composition of the 'target' rocks (shallow sea deposits) and their behaviour under high-pressure shock, shed light on the early phases of the impact and its environmental effects. At Yucatán, the meteorite would have impacted on a sea floor that was naturally rich in water, carbonate, and sulphate; this would have propelled as much as 200 gigatons *each* of sulphur dioxide and water vapour into the stratosphere. Impact models based on the geometry of the crater itself suggest that the impact was oblique and from the south-east. This trajectory would have concentrated the expelled gases towards North America. The fossil record certainly suggests that floral extinctions were particularly severe in this area, but more work elsewhere is needed before this pattern can be verified. The Alvarez and others' work on the effects of the impact suggested that dust and clouds would have plunged the world into a freezing blackout. However, computer modelling of atmospheric conditions now suggests that within a few months light levels and temperatures would have begun to rebound because of the thermal inertia of the oceans, and the steady fall-out of particulate matter from the atmosphere. Unfortunately, however, things would have become no better for some considerable time because the sulphur dioxide and water in the atmosphere would have combined to produce sulphuric acid aerosols, and these would have severely reduced the amount of sunlight reaching the Earth's surface for between five and ten years. These aerosols would have had the combined effects of cooling the Earth to near freezing and drenching its surface in acid rain.

Clearly these estimates are based on computer models, which may be subject to error. However, even if only partly true, the general scope of the combination of environmental effects following the impact would have been genuinely devastating, and may well account for many aspects of the terrestrial and marine extinctions that mark the end of the Cretaceous Period. In a sense, the wonder is that anything survived these apocalyptic conditions.

More perturbations: was it meteorites *and* volcanoes?

While much of the work in recent years has focused on explaining the environmental effects of a large meteorite on global ecosystems, work is still continuing at the Chicxulub site. A major borehole has now been sunk into the crater to a depth of 1.5 km in order that detailed examination of the impact zone can take place. What is beginning to emerge is slightly disturbing to the general pattern that has been explained above. One set of interpretations of the core data indicates that the impact crater may have been made 300,000 years *before* the K-Pg boundary. The interval is represented by 0.5 metres of sediment. This evidence has been used to propose that the end Cretaceous event was not focused on a single large meteorite impact, but several large impacts that occurred right up to the K-Pg boundary—the cumulative effects of which may have resulted in the mass-extinction.

Clearly these new findings indicate that more research and more debate will undoubtedly take place in years to come. Not least among these are the data concerning abundance changes during the approach to the K-Pg boundary and also that concerning massive volcanic activity that coincided with the end Cretaceous events.

Parts of India known as the Deccan represent a series of pulses of flood-basalts that covered large areas of central India; these have been estimated to have had a cumulative volume equivalent to several million cubic kilometres. The global influence of such enormous volcanic eruptions offers another perspective on the events leading up to the K-Pg mass-extinction and has been promoted strongly by Vincent Courtillot and colleagues.

Quite what the environmental impact of such enormous volcanic outpourings was, and, more particularly, whether this was in any way linked to the meteorite impact on the other side of the world seemed doubtful for many years. However recent work has focused on the chemical composition of the lava that forms the Indian Deccan mountain range. Their findings suggest that precisely at the time of the meteorite impact (66 Ma) the composition of the Deccan lava suddenly changed. It is very tempting to attribute this change to the remote effect of the colossal meteorite impact on the other side of the world—perhaps the Earth was the victim of the ultimate 'double-whammy': a simultaneous meteorite impact AND the coincident eruption of huge volumes of lava.

Just contemplating the volcanic event alone (since we have not witnessed a meteorite impact—thank goodness), in the relatively recent past comparatively 'trivial' (on a global scale) volcanic events such as the explosive eruption of Mount St Helens (1980) and in the more distant past the volcanic eruptions that almost completely destroyed the island of Krakatoa (1883) as well as Tambora (1815) are known to have produced measurable climatic change around the world. Quite what the effect of truly enormous volumes of flood basalts emerging on to the Earth's surface might have been can only be guessed at, but reason alone suggests that it would not have been favourable for life.

Mass-extinctions, such as the one that occurred 66 Ma, are truly fascinating punctuation marks in the history of life on Earth—but nailing down exactly what caused them has proved to be, not surprisingly, very difficult given that we are trying to understand, forensically, precisely what happened many tens of millions of years ago.

Dinosaur research now and in the near future

It should be clear by now that a subject such as palaeobiology—certainly as it is currently being applied to fascinating creatures such as dinosaurs—has a decided unpredictability about it. Many research programmes in palaeobiology can be planned, and indeed have an intellectually satisfying structure to them, in order to explore specific issues or problems; this is normal for all the sciences. However, serendipity also plays a significant role in the study of fossils: it can lead research in unexpected directions that could not have been anticipated at the outset. It can also be influenced strongly by spectacular new discoveries—nobody in the early 1990s would have been able to predict the amazing avian-theropod ('dinobird') finds that were made in China in 1996 and continue to the present day. Technological advances in the physical and biological sciences have also played an increasingly important part in research, allowing us to study fossils in ways that were, again, unimaginable just a few years ago.

To take advantage of many of these opportunities it is important to have at hand people who share a number of characteristics. Above all, they need to have an abiding interest in the history of life on Earth as well as naturally inquisitive temperaments. They also need some training in a surprisingly wide range of areas. While it is still important to have individual scientists working and thinking creatively in some degree of isolation, it is increasingly the case that multidisciplinary teams are needed to bring a wider range of skills to bear on each problem, or each new discovery, in order to tease out the information that will move the science a little further forward.

And finally...

My message is a relatively simple one. We, as a human race, could simply choose to ignore the history of life on Earth as it can be interpreted, in part at least, through the study of fossils. There are indeed many who adhere to such a view. Fortunately, I would say, a few of us do not. The pageant of life has been played out across the past ~3,800 million years—a staggeringly long period of time. We as humans currently dominate most ecosystems, either directly or indirectly, but we have only risen to this position of dominance over the past 10,000 years of life on Earth. Before the human species, a wide range of organisms held sway. The dinosaurs were one such group, and they, in a sense, acted as unwitting custodians of the Earth they inhabited. Palaeobiology allows us to trace parts of that custodianship.

The deeper question is: can we learn from past experiences and use them to help us to preserve an inhabitable Earth for other species to inherit when we are finally gone? This is an awesome responsibility given the current global threats posed by exponential population increase, pollution, climatic change, and the threat posed by nuclear power. We are the first species ever to exist on this planet that has been able to appreciate that the Earth is not just 'here and now' but has a deep history. I hope sincerely that we will not also be the last. The one thing that we can be sure of, after studying the waxing and waning of species throughout the immensity of the fossil record, is that the human species will not endure forever.

From our origin as *Homo sapiens* which can be traced back approximately 500,000 years ago, our species might last a further million years, or perhaps even five million years if we are extraordinarily successful (or lucky), but we will eventually go the way of the dinosaurs: that much at least is written in the rocks.

Further reading

Introduction

D. R. Dean, *Gideon Mantell and the Discovery of Dinosaurs* (Cambridge: Cambridge University Press, 1999).

A. J. Desmond, *The Hot-Blooded Dinosaurs: A Revolution in Palaeontology* (London: Blond & Briggs, 1975).

A. Mayor, *The First Fossil Hunters: Palaeontology in Greek and Roman Times* (Princeton: Princeton University Press, 2001).

C. McGowan, *The Dragon Seekers* (Cambridge, MA: Perseus Publishing, 2001).

D. B. Norman, *The Illustrated Encyclopedia of Dinosaurs* (London: Salamander Books, 1985).

D. B. Norman, *Dinosaur!* (London: Boxtree, 1991).

D. B. Norman, *Dinosaurs: A Very Short Introduction* (1st edn, Oxford: Oxford University Press, 2005).

M. J. S. Rudwick, *The Meaning of Fossils: Episodes in the History of Palaeontology* (New York: Science History Books, 1976).

N. A. Rupke, *Richard Owen: Victorian Naturalist* (New Haven: Yale University Press, 1994).

Chapter 1: Dinosaurs in perspective

D. E. G. Briggs and P. R. Crowther (eds), *Palaeobiology II* (Oxford: Blackwell Science, 2001).

E. Casier, *Les Iguanodons de Bernissart* (Brussels: Institute Royal des Sciences Naturelles de Belgique, 1978).

B. Charlesworth and D. Charlesworth, *Evolution: A Very Short Introduction* (Oxford: Oxford University Press, 2003).

E. H. Colbert, *Men and Dinosaurs: The Search in Field and Laboratory* (London: Evans Brothers Ltd, 1968).

C. R. Darwin, *On the Origin of Species by Means of Natural Selection, or the Preservation of Favoured Races in the Struggle for Life* (London: John Murray, 1859).

A. J. Desmond, *The Hot-Blooded Dinosaurs: A Revolution in Palaeontology* (London: Blond & Briggs, 1975).

R. A. Fortey, *The Earth: An Intimate Story* (London: Harper Perennial, 2005).

D. B. Norman, 'On the history of the discovery of fossils at Bernissart in Belgium', *Archives of Natural History* 14 (1987): 59–75.

D. B. Norman, 'Gideon Mantell's Mantel-piece: the earliest well-preserved ornithischian dinosaur', *Modern Geology* 18 (2001): 225–45.

D. B. Norman, *Dinosaurs: A Very Short Introduction* (1st edn, Oxford: Oxford University Press, 2005).

R. Owen, 'Report on British Fossil Reptiles. Part 2', *Report of the British Association for the Advancement of Science (Plymouth)* 11 (1841): 60–204.

Chapter 2: Dinosaurs updated

R. T. Bakker, 'Dinosaur renaissance', *Scientific American* 475 (1975): 58–78.

S. L. Brusatte, *Dinosaur Paleobiology* (London: John Wiley and Sons, 2012).

G. Heilmann, *The Origin of Birds* (New York: D. Appleton, 1927).

D. B. Norman, *Dinosaur!* (London: Boxtree, 1991).

J. H. Ostrom, *Osteology of* Deinonychus antirrhopus, *an Unusual Theropod from the Lower Cretaceous of Montana* (New Haven: Peabody Museum of Natural History, Yale University, Bulletin 30, 1969), 172pp.

J. H. Ostrom and J. S. McIntosh, *Marsh's Dinosaurs: The Collections from Como Bluff* (New Haven: Yale University Press, 1966).

Chapter 3: A new perspective on *Iguanodon*

M. D. Brasier, D. B. Norman, et al., 'Remarkable preservation of brain tissues in an Early Cretaceous iguanodontian dinosaur'. In A. T. Brasier, D. McIlroy, and N. McLoughlin (eds), *Earth*

System Evolution and Early Life (London: The Geological Society of London Special Publications, 2017), 448.

D. B. Norman, 'On the history of the discovery of fossils at Bernissart in Belgium', *Archives of Natural History* 14 (1987): 59–75.

D. B. Norman, 'On the ornithischian dinosaur *Iguanodon bernissartensis* from Belgium', *Mémoires de l'Institut Royal des Sciences Naturelles de Belgique* 178 (1980): 105pp.

D. B. Norman, 'On the anatomy of *Iguanodon [=Mantellisaurus] atherfieldensis* (Ornithischia: Ornithopoda)', *Bulletin de l'Institut Royal des Sciences Naturelles de Belgique, Sciences de la Terre* 56 (1986): 281–372.

D. B. Norman, 'Basal Iguanodontia'. In D. B. Weishampel, P. Dodson, and H. Osmolska (eds), *The Dinosauria* (Berkeley, CA: California University Press, 2004), 413–37.

D. B. Norman, *Very Short Introduction: Dinosaurs* (1st edn, Oxford: Oxford University Press, 2005).

D. B. Norman and D. B. Weishampel, 'Ornithopod feeding mechanisms: their bearing on the evolution of herbivory', *American Naturalist* 126 (1985): 151–64.

Chapter 4: Relationships between dinosaurs

M. G. Baron, D. B. Norman, and P. M. Barrett, 'A novel hypothesis of dinosaur relationships and early dinosaur evolution', *Nature* 543 (2017): 501–6.

S. L. Brusatte, *Dinosaur Paleobiology* (London: John Wiley and Sons, 2012).

R. A. Fortey, *The Earth: an Intimate Story* (London: Harper Perennial, 2005).

J. Gauthier, 'Saurischian monophyly and the origin of birds'. In K. Padian (ed.), *The Origin of Birds and the Evolution of Flight: Memoirs of the California Academy of Science* 8 (1986): 1–55.

T. H. Huxley, 'On the classification of the Dinosauria with observations on the Dinosauria of the Trias', *Quarterly Journal of the Geological Society of London* 26 (1870): 32–51.

D. B. Norman, 'A systematic reappraisal of the reptile order Ornithischia'. In W. E. Reif and F. Westphal (eds), *Third Symposium on Mesozoic Terrestrial Ecosystems: Short Papers* (Tubingen: Attempto Verlag, 1984), 157–62.

D. B. Norman, 'On Mongolian Ornithopods (Dinosauria: Ornithischia). 1. Iguanodon orientalis Rozhdestvensky 1952', *Zoological Journal of the Linnean Society* 116 (1996): 303–15.

H. G. Seeley, 'On the classification of the fossil animals commonly named Dinosauria', *Proceedings of the Royal Society of London* 43 (1887): 165–71.

D. B. Weishampel, Z. Csiki, C. Jianu, and D. B. Norman, 'Osteology and phylogeny of Zalmoxes (n. g.), an unusual euornithopod dinosaur from the latest Cretaceous of Romania', *Journal of Systematic Palaeontology* 1(2) (2003): 1–56.

P. Upchurch, C. Hunn, and D. B. Norman, 'An analysis of dinosaurian biogeography: evidence for the existence of vicariance and dispersal patterns caused by geological events', *Proceedings of the Royal Society B* 269 (2002): 613–21.

A. Wegener, *The Origins of the Continents and Oceans* (4th edn, London: Methuen and Co. Ltd, 1966).

Chapter 5: Dinosaurs: their biology and way of life

D. E. G. Briggs and P. R. Crowther (eds), *Palaeobiology II* (Oxford: Blackwell Science, 2001).

S. L. Brusatte, *Dinosaur Paleobiology* (London: John Wiley and Sons, 2012).

C. Lavers, *Why Elephants Have Big Ears* (London: Gollancz, 2000).

Q. Li et al., 'Melanosome evolution indicates a key physiological shift within feathered dinosaurs', *Nature* 507 (2014): 350–4.

C. McGowan, *Dinosaurs, Spitfires & Sea Dragons* (Cambridge, MA: Harvard University Press, 1992).

D. B. Norman, *Dinosaur!* (London: Boxtree, 1991).

D. B. Norman, *Prehistoric Life: The Rise of the Vertebrates* (London: Boxtree, 1994).

D. B. Norman and P. Wellnhofer, *The Illustrated Encyclopedia of Dinosaurs* (London: Salamander Books, 2000).

P. M. Sander et al., 'Biology of the sauropod dinosaurs: the evolution of gigantism', *Biological Reviews* 86 (2011): 117–55.

K. Stein and E. Prondvai, 'Rethinking the nature of fibrolamellar bone: an integrated biological revision of sauropod plexiform bone formation', *Biological Review* 89 (2014): 24–47.

R. D. K. Thomas and E. C. Olson, *A Cold Look at the Warm-Blooded Dinosaurs* (Boulder, CO: Westview Press, 1980).

D. B. Weishampel, P. Dodson, et al. (eds), *The Dinosauria* (Berkeley, CA: University of California Press, 2004).

X. Xu et al., 'An integrative approach to the understanding of bird origins', *Science* 346 (2014): 1341.

E. Yong, 'Dust-up over dinosaurs' true colours', *Nature News*, 27 March 2013.

Chapter 6: Dinosaur research: observations and techniques

S. L. Brusatte, *Dinosaur Paleobiology* (London: John Wiley and Sons, 2012).

R. De Salle and D. Lindley, *The Science of Jurassic Park and the Lost World, or How to Build a Dinosaur* (London: Harper Collins, 1997).

R. A. Eagle et al., 'Dinosaur body temperature determined from isotopic (^{13}C-^{18}O) ordering in fossil biominerals', *Science* 333 (2011): 443–5.

D. C. Evans et al., Endocranial anatomy of Lambeosaurine Hadrosaurids (Dinosauria: Ornithischia): a sensorineural perspective on Cranial Crest Function', *The Anatomical Record* 292 (2009): 1315–37.

J. R. Hutchinson et al., 'A computational analysis of limb and body dimensions in Tyrannosaurus rex with implications for locomotion, ontogeny and growth', *PloS ONE* (doi.org/10.1371/journal.pone.0026037, 2011).

M. G. Lockley, *Tracking Dinosaurs: A New Look at an Ancient World* (Cambridge: Cambridge University Press, 1991).

M. G. Lockley et al., 'Theropod courtship: large scale physical evidence of display arenas and avian-like scrape ceremony behaviour by Cretaceous dinosaurs', *Nature Scientific Reports* 6 (2015): 1895–2.

D. B. Norman, *Dinosaurs: A Very Short Introduction* (1st edn, Oxford: Oxford University Press, 2005).

E. J. Rayfield, D. B. Norman, et al., 'Cranial design and function in a large theropod dinosaur, *Nature* 409 (2001): 1033–7.

M. H. Schweitzer et al., 'Heme compounds in dinosaur trabecular bone', *Proceedings of the North American Academy of Sciences* 94 (1994): 6291–6.

M. H. Schweitzer et al., 'Soft-tissue vessels and cellular preservation in Tyrannosaurus rex', *Science* 307 (2005): 1952–5.

M. H. Schweitzer et al., 'Analyses of soft tissue from Tyrannosaurus rex suggest the presence of protein', *Science* 316 (2007): 277–80.

J. M. Asara et al., 'Protein sequences from Mastodon and Tyrannosaurus rex revealed by mass spectrometry', *Science* 316 (2007): 280–5.

C. Zimmer, 'Is dinosaur "soft tissue" really slime?', *Science* 321 (2008): 623.

R. M. Service, 'Protein in 80 million year old fossil bolsters controversial T. rex claim', *Science* 324 (2009): 578.

Chapter 7: The future of research on the past

L. W. Alvarez, W. Alvarez, F. Asaro, and H. V. Michel, 'Extraterrestrial cause for the Cretaceous-Tertiary Extinction', *Science* 208 (1980): 1095–108.

R. B. J. Benson et al., 'Rates of dinosaur body mass evolution indicate 170 million years of sustained ecological innovation on the avian stem lineage', *PloS Biol* 12(5) (2015): e10011853.

S. L. Brusatte et al., 'The extinction of the dinosaurs', *Biological Reviews* 90 (2015): 628–4.

R. J. Butler et al., 'Sea level, dinosaur diversity and sampling biases: investigating the "common cause" hypothesis in the terrestrial realm', *Proceedings of the Royal Society B* 278 (2011): 1165–70.

V. Courtillot, *Evolutionary Catastrophes: The Science of Mass Extinctions* (Cambridge: Cambridge University Press, 1999).

G. Keller, A. Sahni, and S. Bajpai, 'Deccan vulcanism, the KT mass extinction and dinosaurs', *Journal of Bioscience* 24 (2009): 709–28.

P. R. Renne et al., 'State shift in Deccan volcanism at the Cretaceous-Paleogene boundary, possibly induced by impact', *Science* 350(6256) (2015): 76–8.

M. Sakamoto, M. J. Benton, and C. Venditti, 'Dinosaurs in decline tens of millions of years before their final extinction', *Proceedings of the North American Academy of Sciences* 113 (2016): 5036–40.

Index

Dinosaurs

SOCIAL MEDIA
Very Short Introduction

Join our community

www.oup.com/vsi

- Join us online at the official Very Short Introductions **Facebook** page.
- Access the thoughts and musings of our authors with our online **blog**.
- Sign up for our monthly **e-newsletter** to receive information on all new titles publishing that month.
- Browse the full range of Very Short Introductions online.
- Read **extracts** from the Introductions for free.
- If you are a teacher or lecturer you can order inspection copies quickly and simply via our website.

ONLINE CATALOGUE
A Very Short Introduction

Our online catalogue is designed to make it easy to find your ideal Very Short Introduction. View the entire collection by subject area, watch author videos, read sample chapters, and download reading guides.

http://global.oup.com/uk/academic/general/vsi_list/

GLOBALIZATION
A Very Short Introduction
Manfred Steger

'Globalization' has become one of the defining buzzwords of our time - a term that describes a variety of accelerating economic, political, cultural, ideological, and environmental processes that are rapidly altering our experience of the world. It is by its nature a dynamic topic - and this *Very Short Introduction* has been fully updated for 2009, to include developments in global politics, the impact of terrorism, and environmental issues. Presenting globalization in accessible language as a multifaceted process encompassing global, regional, and local aspects of social life, Manfred B. Steger looks at its causes and effects, examines whether it is a new phenomenon, and explores the question of whether, ultimately, globalization is a good or a bad thing.

www.oup.com/vsi

CANCER
A Very Short Introduction
Nick James

Cancer research is a major economic activity. There are constant improvements in treatment techniques that result in better cure rates and increased quality and quantity of life for those with the disease, yet stories of breakthroughs in a cure for cancer are often in the media. In this *Very Short Introduction* Nick James, founder of the CancerHelp UK website, examines the trends in diagnosis and treatment of the disease, as well as its economic consequences. Asking what cancer is and what causes it, he considers issues surrounding expensive drug development, what can be done to reduce the risk of developing cancer, and the use of complementary and alternative therapies.